Springer Optimization and Its Applications

VOLUME 72

Managing Editor
Panos M. Pardalos (University of Florida)

Editor–Combinatorial Optimization
Ding-Zhu Du (University of Texas at Dallas)

T0214300

Advisory Board
J. Birge (University of Chicago)
C.A. Floudas (Princeton University)
F. Giannessi (University of Pisa)
H.D. Sherali (Virginia Polytechnic and State University)
T. Terlaky (McMaster University)
Y.Ye (Stanford University)

Aims and Scope
Optimization has been expanding in all directions at an astonishing rate during the last few decades. New algorithmic and theoretical techniques have been developed, the diffusion into other disciplines has proceeded at a rapid pace, and our knowledge of all aspects of the field has grown even more profound. At the same time, one of the most striking trends in optimization is the constantly increasing emphasis on the interdisciplinary nature of the field. Optimization has been a basic tool in all areas of applied mathematics, engineering, medicine, economics, and other sciences.

The series *Springer Optimization and Its Applications* publishes under-graduate and graduate textbooks, monographs, and state-of-the-art expository work that focus on algorithms for solving optimization problems and also study applications involving such problems. Some of the topics covered include nonlinear optimization (convex and nonconvex), network flow problems, stochastic optimization, optimal control, discrete optimization, multiobjective programming, description of software packages, approximation techniques and heuristic approaches.

For further volumes:
http://www.springer.com/series/7393

Ivan V. Sergienko

Methods of Optimization and Systems Analysis for Problems of Transcomputational Complexity

 Springer

Ivan V. Sergienko
National Academy of Sciences of Ukraine
V.M. Glushkov Institute of Cybernetics
Kiev, Ukraine

ISSN 1931-6828
ISBN 978-1-4939-0023-7   ISBN 978-1-4614-4211-0 (eBook)
DOI 10.1007/978-1-4614-4211-0
Springer New York Heidelberg Dordrecht London

Printed on acid-free paper

Springer is part of Springer Science+Business Media (www.springer.com)

V. M. Glushkov Institute of Cybernetics of the National Academy of Sciences of Ukraine. Main building

Academician V. S. Mikhalevich

Preface

This book is devoted to the elucidation of scientific achievements of academician Volodymyr Sergiyovych Mikhalevich, a well-known Ukrainian scientist and educator, and his role in the development of modern informatics. His 80th anniversary is an opportunity to think over his life, his strenuous work in science, and the influence of his ideas and personality on the becoming and development of the work to which he dedicated his life. The author was fortunate to work with V. S. Mikhalevich over a period of three decades and to witness his devotion to science as well as his vision for the future of Ukrainian informatics and mathematics. He was a talented scientist and a great organizer of science, a benevolent and generous teacher, and a kind-hearted and sensitive man. The combination of these features helped him to establish himself not only as a scientist but also as the founder of the world's well-known Ukrainian school of optimization theory. Professor Mikhalevich had many disciples who became worthy continuers of his noble cause. It is precisely scientific directions and some results of activity of Mikhalevich and representatives of his school that are considered in this book.

The author did not set himself the task of covering all lines of investigations (all the more, all works) of Mikhalevich. The list of his publications consists of several hundred works. It would be desirable to mention only the main (from the viewpoint of the author) works. For this reason, this work contains references (as required) to a part of the works of Mikhalevich and his disciples and colleagues. A comprehensive analysis of the scientific work of Mikhalevich, undoubtedly, deserves a separate exploration.

Professor Mikhalevich expired before his 65th birthday and did not realize many of his plans, but his achievements left an appreciable mark in science. It is necessary to emphasize that Mikhalevich could often determine prospects of scientific ideas and the path of development of some scientific direction or other, and could skillfully formulate mathematical statements of main problems better than anyone else. His high mathematical culture manifested itself through all his activities during the consideration of questions of formalization (adequate mathematical description) of complicated processes, choice of effective approaches to the solution of problems that arose in this case, scientific significance of the results obtained, etc. He became the measure of a scientist not only to his colleagues (the author considers V. S. Mikhalevich to be one of his teachers in the field of optimization theory) but also for his numerous followers in Ukraine and abroad.

Mikhalevich had to work in times that were hard for cybernetics. He began his scientific career when the new science had not yet been recognized and accepted by the society. Only a few people managed to immediately understand and appreciate its importance for human development. Mikhalevich worked with academician V. M. Glushkov for a long time and was his staunch "comrade-in-arms" and one of the organizers of the development of cybernetics not only in Ukraine but also in the USSR as a whole. After the Glushkov's death, Mikhalevich headed the Institute of Cybernetics of the National Academy of Sciences (NAS), Ukraine, for 12 years and made a considerable contribution to the development of cybernetic investigations and to the organization of professional training in cybernetics in higher education institutions of Ukraine. This earned him high appreciation by the society, and he was permanently supported by the presidium of NAS of Ukraine and well-known experts and organizers of science in the USSR such as academicians A. A. Dorodnitsyn, N. N. Bogolyubov, and M. V. Keldysh. This support helped to develop the Institute of Cybernetics (and the Cybernetic Center of NAS of Ukraine later on) and to organize the work on informatization in the country as a whole.

Nowadays, we are frequently faced with the following question: What scientific directions are topical such that they will have an impact on the development of our economy and scientific research in the near future? It is not easy to answer. Nevertheless, it becomes more and more obvious that one cannot do without full-scale informatization in this case. In particular, Barack H. Obama, the current president of the United States, emphasized the necessity of the active development of informatics as one of three most important scientific directions in the modern era at a meeting with scientists (at the very beginning of his presidency). We who work in this field always felt its importance to the life of society and the responsibility for it. This was bequeathed to us by our great preceptors V. M. Glushkov, A. A. Dorodnitsyn, and V. S. Mikhalevich who felt (better than anyone else) prospects for progress and saw the road of progress in the distant future.

In this work, the main accent is placed on the solution of problems of transcomputational complexity, i.e., problems that, within the framework of a definite model of calculations and some fixed software of a computer system, are characterized by ultrahigh values of complexity estimates. As a rule, mathematical formulations of these problems use 100,000 (and even ten million) variables and approximately the same number of various constraints. Examples of such problems are presented.

This work consists of five chapters in which scientific achievements of V. S. Mikhalevich and his disciples in the field of optimization theory, economic cybernetics, and construction of computer complexes and their software, in particular, packages of applied programs, are analyzed. Special attention is given to the construction of optimization methods for the solution of complicated problems in various statements. Problems of linear programming, problems of stochastic and discrete optimization, and also (partially) problems of integer programming are considered. By their very nature, such problems, as a rule, have transcomputational complexity. It is obvious that the mentioned problems can be efficiently solved on supercomputers that have ultrahigh computational speed (this speed sometimes amounts to 100 and 1,000 billion operations per second) and corresponding memory sizes.

It is precisely the promising investigations in informatics with the direct participation of V. S. Mikhalevich that the author tries to stress, namely, the research and development activities that were initiated by him or some of these activities that were carried out under his scientific and methodical leadership. However, a considerable part of the works and results that are considered in this book have been accomplished by his disciples and followers already in recent years (1995–2010), i.e., without the participation of Mikhalevich.

In preparing this work for publication, the author has used materials and recommendations of many disciples and colleagues of V. S. Mikhalevich. Taking the opportunity, the author expresses his sincere gratitude to all of them.

Acknowledgments

The author wishes to thank V.A. Grebnev and O.S. Nikonova for translating this book into English and to G.V. Zorko for composition of the manuscript.

The author is also grateful to Elizabeth Loew, Senior Editor, and to Jacob Gallay, Assistant Editor, for their helpful support and collaboration in preparation of the manuscript.

Contents

Chapter 1
Science Was the Meaning of His Life

Abstract This chapter gives a general description of the life and scientific career of V. S. Mikhalevich and his scientific and public activities. The development of one of the first (in the USSR) supercomputer systems under the direction of V. M. Glushkov is considered on which, already under the direction of V. S. Mikhalevich, complicated problems were solved by scientists of the Institute of Cybernetics. It was then (in the 1980s) that this system underlay the beginning of obtaining the experience in the parallelization of computational processes, and pioneer works were performed on the creation of software for supercomputer systems capable of realizing parallel processes. These works undoubtedly exerted a positive influence on the development of the well-known supercomputers of the SCIT family and also supercomputers of the INPARCOM family developed by scientists of the Institute of Cybernetics together with specialists of the State Scientific-Production Enterprise (SSPE) "Elektronmash."

1.1 Beginning: Education and the Path to Big Science

For us, the year 2009 has passed under the sign of the star of N. N. Bogolyubov, a man of genius; his centenary was celebrated everywhere, and these conferences, meetings, and recollections taken together induced meditations and made it possible not to simply realize but also to see the linkage of times, a continuous wave of discipleship and teaching, its flow from mind to mind, from one person to another, and, eventually, from heart to heart.

It was precisely N. N. Bogolyubov who introduced mathematical circles for talented schoolchildren of the seventh–tenth grades at the Taras Shevchenko State University (National University after 1994) of Kyiv (TSSUK) since he understood very well the importance of the initiation into a source of knowledge in due time and who himself formerly had attentive and generous teachers.

I.V. Sergienko, *Methods of Optimization and Systems Analysis for Problems of Transcomputational Complexity*, Springer Optimization and Its Applications 72, DOI 10.1007/978-1-4614-4211-0_1, © Springer Science+Business Media New York 2012

The classes were held in the Red University Building of the TSSUK, and each of them consisted of two lessons devoted to algebra and geometry. From time to time, professors of the Faculty of Mechanics and Mathematics of the university delivered general lectures. At the end of April, a mathematical olympiad was organized.

These circles were conducted by second-year or older students of the faculty. Many students who subsequently became famous scientists, academicians, and also excellent maths masters owing to whom the number of entrants to the faculty does not decrease until now went through this school of entry-level teaching.

One of the students who conducted the lesson on circles was V. Mikhalevich, and it was his lessons which were actively attended by Kyiv senior schoolchildren and, among them, I. Kovalenko and A. Letichevskii who are academicians now and whose works are well known in the scientific world. And they are not only just well known but their results are widely used in practice. Academician Kovalenko calls V. Mikhalevich one of his chief teachers. He himself has trained many disciples and, hence, it may be said that the wave of teaching–schooling reliably moves and, hopefully, will not discontinue and will not disappear in the quicksand of time.

At the university, the teacher of V. Mikhalevich (and also I. Kovalenko later on) was B. V. Gnedenko, the founder and leader of the Ukrainian School of Probability Theory, which is well known all over the world. After graduating from the University in 1953, V. Mikhalevich became a postgraduate student of B. V. Gnedenko. But, later on, V. Mikhalevich together with V. Korolyuk and A. Skorohod, two other postgraduate students of B. V. Gnedenko (who went on a long foreign academic mission), went on an academic trip to the Lomonosov Moscow State University at which the world-known school headed by Academician A. N. Kolmogorov functioned at that time. V. S. Mikhalevich became a favorite disciple of A. N. Kolmogorov.

Two years later, they came back to Kyiv and soon became candidates of sciences. They lectured at the Faculty of Mechanics and Mathematics of the Taras Shevchenko State University of Kyiv. At that time, the author was a second-year student and listened to a special course of V. S. Mikhalevich.

Talking about his "chief teacher," I. N. Kovalenko mentions the happy times for him and his coevals when V. Mikhalevich was not yet absorbed in enormous amount of organizational work that swallowed the greatest part of his time; at that time, he still had time to give day-to-day advice and leadership to young researchers in a great many lines of investigations. They consisted of various aspects of optimization for N. Shor and Yu. Ermoliev, software tools of simulation for T. Mar'yanovich, sequential pattern recognition methods for V. Kovalevskii, robust criteria for P. Knopov, network planning methods for V. Shkurba, and statistical decision functions for I. Kovalenko.

Volodymyr Mikhalevich, a student

A great many persons can testify to the immense prestige of V. S. Mikhalevich among researchers of military operations. The mathematical methods created by V. S. Mikhalevich and his disciples had significant repercussions in the military engineer community and were applied to the solution of many optimization problems of planning and maintenance service.

V. Mikhalevich had a good temper, could mix with people easily, and was well-wishing and communicative. He was handsome and looked elegant and fashionable even in a very ordinary suit. From time to time, he seemed to be carefree, a lucky man who had not faced neither difficulties nor failures. If something darkened his life, one could not guess this by the look of his face.

In fact, as was clarified later when we already worked together, he had difficult times. They were, probably, as those times of the majority of us whose childhood was broken by war. He was born on March 10, 1930, and, hence, was an 11-year-old boy in 1941. The family lived at the center of Chernihiv in a former bishop house divided into several walk-in apartments during the years of Soviet Power. The house stood in a garden.

The childhood days came to an end when the war came to the city. One day, that is, August 23, 1941, stuck in his memory for the rest of his life. Fascists made the decision to illustrate their military power by the example of Chernihiv, to wipe it off the face of the earth, and to kill or to terrify all to death. After 3 days, his family left the city, and they became refugees. They went to Stalingrad, where his native aunt (a sister of his mother) lived, and stayed at her place for a year. He lived among relatives, and things were seemingly going rather well except that Volodya had typhus.

But we should not forget that he was in Stalingrad. Exactly 1 year later, on the same date, that is, on August 23, 1942, a similar violent attack on the city took place. Even after many years and after going through tough times, he said that those two August days were most terrible in his life. Volodya just came home from the hospital. Street battles were being fought in the city. At the beginning of September, they were aboard on one of the last steamships that sailed down the Volga toward Saratov. They were soon landed in the frontline area since his mother fell ill with typhus. His father was at the front. It was only he, a 12-year-old boy, who could earn a living. He carted milk to a milk plant on a dray. And he also had to go to school.

After coming back to their destructed native city, they were amazed to find that their bishop house remained intact. And the verdant garden looked like before. But all the surroundings were in complete ruins. Despite the trials and troubles of his life, Volodya finished school at 17 in 1947, was awarded his high-school diploma and a silver medal, and became the student of the Faculty of Mechanics and Mathematics of the Taras Shevchenko State University in the same summer.

Even in his studentship, V. Mikhalevich published three scientific articles (two of them with coauthors) in "Doklady Akad Nauk SSSR" ("Proceedings of the USSR Academy of Sciences"). Science became the meaning of his life.

The intensive development of mass production in postwar years created the acute problem of elaboration of a statistical sampling procedure that allows one, based on the results of checking a limited number of manufactured articles, to draw a conclusion on the quality of their high-production run. In this case, an increase in the check sample size decreases the risk of acceptance of a spoiled production lot but predetermines an increase in the cost of statistical experiments. The problem arose of searching for an optimal ratio between the cost of a sample and possible losses from the acceptance of a spoiled production lot. It was such problems that were the subject of investigations of V. S. Mikhalevich. He proposed original acceptance sampling methods. These results underlay the Ph.D. thesis defended by him.

Since the application of decision-making methods proposed by him and based on a sequential exhaustive search for variants was connected with a considerable amount of computation, V. S. Mikhalevich took an interest in the possibility of using the computer MESM (Small Electronic Calculating Machine) for this purpose. This computer was actively exploited in the Laboratory of Computer Engineering at the Institute of Mathematics of the Academy of Sciences of the UkrSSR at that time. Together with his colleagues V. Korolyuk and E. Yushchenko, he got involved in investigations carried out with the use of the computer.

At the same time, B. V. Gnedenko, the director of the Institute of Mathematics, offered V. M. Glushkov to head the Laboratory of Computer Engineering whose themes were gradually extended. And works started that were directed toward the creation of new models of computing facilities.

After the creation of the Computer Center in 1957, V. M. Glushkov who became its director offered V. S. Mikhalevich to head a group of specialists with a view to pursuing investigations in the field of reliability of electronic circuits. In this manner, his activity began in the field of informatics. It was suggested that the use of computing machinery for the management of the national economy would open new great prospects. In this connection, the Department of Automated Industrial Management Systems was organized at the Computer Center in 1960. Mikhalevich was elected its head. Later on, this department became the Department of Economic Cybernetics whose main lines of investigations were problems of optimal planning, operations research, design of complicated objects, and systems of automation of processes in production and transport spheres. At the department, simultaneously with pursuing scientific investigations, the work on professional training was placed on a wide footing. Only during 3 years (1960–1962), over a hundred specialists from different regions of the Soviet Union were trained here.

Developing algorithms for numerical solution of extremal problems of technical and economic planning, V. S. Mikhalevich paid attention to the expediency of using ideas of the theory of sequential statistical decisions. As a result, the scheme of sequential analysis of variants was substantiated and numerical algorithms for its realization on computers were proposed. Mikhalevich made a report on these results at the Fourth All-Union Mathematical Congress in 1961.

The method of sequential analysis of variants was very quickly and widely recognized and applied. At the suggestion of Academician N. N. Moiseev, Muscovites called this method the "Kyiv broom." The "Kyiv broom" became one of the main tools in solving problems of optimal design of road networks and electric and gas networks; in finding shortest paths in graphs and critical paths in problems of network planning; in distribution of industry, scheduling theory, and scheduling; and in solving many other problems.

The idea of the "Kyiv broom" method gave impetus to the creation of some other algorithmic schemes and methods for the solution of complicated optimization problems. In particular, they included extremely complicated discrete optimization problems and stochastic programming problems. Based on these ideas, the Kyiv school of optimization was formed with which were affiliated tens of collectives of the National Academy of Sciences (NAS) of Ukraine; universities of Kyiv, Dnipropetrovsk, Zaporizhia, Kharkiv, Lviv, and Uzhgorod; National Technical University of Ukraine "Kyiv Polytechnic Institute" (NTUU "KPI"); and a lot of design establishments and organizations. The school includes leading specialists in various fields of science.

As is well known, active scientific investigations in the field of optimization and control were launched in Europe and in the USA in the 1960s. Our scientists also were not out of the swim. Academician N. N. Moiseev, the well-known Russian scientist, organized periodically functioning summer schools on optimization for specialists of this profile. First of them took place in Chernivtsi (Ukraine), Shemas (Azerbaijan), and Tiraspol (Moldova). Over a long period of time, N. N. Moiseev played the role of the curator of investigations into optimization in the USSR. In Ukraine, this role was played by V. S. Mikhalevich who had considerable achievements in this field.

Here, it is also necessary to bring to mind that republican summer schools on computational mathematics and theory of computation are regularly held in Ukraine over more than 30 years (the 35th school took place in Katsiveli in 2009). These schools gradually became international. V. M. Glushkov and V. S. Mikhalevich took an active part in their work at that time together with other leading specialists in the theory of computations from the Institute of Cybernetics, which was and remains the organizer of these schools. Scientists of Russia, Belarus, Moldova, and other countries regularly participate in the work of summer schools. It is very important that this work was accompanied by issuing scientific works of lecturers of this school. V. S. Mikhalevich was the research supervisor of summer schools for many years. It is pertinent to note that the active work in carrying out these schools was performed by well-known scientists such as V. V. Ivanov, V. K. Zadiraka, M. D. Babich, I. I. Lyashko, B. M. Bublik, B. N. Pshenichnyi, Yu. M. Yermoliev, N. Z. Shor, V. S. Deineka, V. V. Skopetskii, A. A. Chikrii, P. I. Andon, A. A. Letichevskii, I. N. Molchanov, Yu. G. Krivonos, and many other specialists who work in the system of the National Academy of Sciences of Ukraine and in institutions of higher education of Ukraine.

Based on the scheme of sequential analysis of variants, V. S. Mikhalevich and N. Z. Shor proposed a general optimality principle for monotonically recursive functionals that is a rather wide generalization of Bellman's optimality principle in dynamic programming. We will dwell in more detail on these questions and on many generalizations of this algorithmic scheme in Chap. 2.

At a conference in Katsiveli. Standing: P. I. Andon, N. F. Kaspshitskaya, O. S. Stukalo, and N. I. Tukalevskaya. Sitting: I. V. Sergienko and V. S. Mikhalevich

In 1963, V. S. Mikhalevich was designated the coordinator of works on the implementation of systems of network planning and management in main machine-building and defense industries of the USSR and also in building industry. This work facilitated the use of systems for the calculation of network diagrams for managing many large design projects and construction of important objects of nationwide significance.

Problems of optimal scheduling and network problems of optimal distribution of limited resources are special problems of discrete, in particular, integer-valued or Boolean programming. After the creation of the theory of computational complexity of extreme combinatorial problems, it became obvious that the absolute majority of them belong to the class of *NP*-complete problems of computational mathematics. A practical deduction drawn from this theory is that the existence of rather simple efficient computational algorithms for these problems is very improbable. Therefore, only very complicated computational schemes for searching for optimal decisions (an example of such a scheme is the sequential analysis of variants) remain at our disposal. First attempts of describing algorithms of sequential analysis of variants for scheduling problems in some more or less simple statement (of the type "machine tools-live parts") belong to V. S. Mikhalevich and V. V. Shkurba. The merit of V. V. Shkurba is also the use of the idea of the method of sequential analysis of variants to solve many applied problems of scheduling and production management.

As is well known, network problems of optimal distribution of limited resources, which are formulated in terms of integer-valued and Boolean programming, have a specificity that makes it possible to distinguish between them and other problems. Their constraint matrices are composed of the following two parts: the so-called general resource constraints and network constraints that characterize acyclic graphs. Despite the computational complexity of problems, it is this specificity that allows one to use the entire arsenal of accumulated methods of mathematical programming for searching for computational methods of increased computational efficiency.

In 1966, V. S. Mikhalevich organized the First All-Union Conference on Mathematical Problems of Network Planning and Management. The conference consolidated many specialists of the country in solving problems of optimum management of the national economy and outlined prospects of development of this line of investigation. At the same time, considerable scientific forces were formed at the Institute of Cybernetics that were able to independently perform demanding works and to develop urgent lines of investigations in the cybernetic science. In the 1960s, fundamentally new results were obtained and efficient numerical algorithms were proposed; in particular, the generalized gradient descent method was proposed that makes it possible to minimize convex functions with discontinuous gradients, that is, can be used in nonsmooth optimization. Later on, the generalized gradient descent method was extended to the class of convex programming problems. By the way, similar methods were proposed in Western countries almost 10 years later when Western researchers had understood that they are key methods for the solution of high-dimensional problems. Later on, N. Z. Shor developed this scientific line of investigation. He proposed and experimentally investigated a subgradient-type method with space dilation destined for the optimization of multiextremal functions. This method remains the most efficient procedure of nonsmooth optimization until now. It practically covered the ellipsoid method that was proposed later on by our and foreign specialists and was well known and well advertised in the world. At the 11th International Symposium on Mathematical Programming Problems that took place in 1982, V. S. Mikhalevich and N. Z. Shor presented a report on the development of methods and technologies of solution of high-dimensional problems. This report aroused considerable interest in the community of scientists.

In [95], which sums up this line of investigation, V. S. Mikhalevich and A. I. Kuksa developed and theoretically and experimentally substantiated a series of approaches to the solution of problems of this type. Here, we are dealing with the classical understanding of dynamic programming (of R. Bellman) and sequential analysis of variants and also with a developed scheme of the branch-and-bound method. New schemes of usage and optimization of the so-called dual estimates in the branch-and-bound method that are considered in [95] relate them with one more line of investigation mentioned above, namely, the generalized gradient descent method for large-dimensional problems.

A number of fundamental results on the development of integer-valued linear programming that were published in [116] later on were used in solving important classes of extremal problems on graphs (problems of optimal distribution, covering, packing, design of communication networks, etc.).

One more important scientific line of investigation of V. S. Mikhalevich together with his disciples became the development of stochastic optimization methods that take into account the probabilistic nature of processes being investigated and the risk connected with indefiniteness, which is a characteristic feature of a decision-making process. These methods can be applied to the solution of complicated applied problems in which the object being investigated quickly and unexpectedly changes its behavior. This line of investigation was developed in works of Yu. M. Yermoliev and his disciples, which will also be considered in Chap. 2.

Among many applied lines of investigations that were personally developed (or even initiated) by V. S. Mikhalevich, it makes sense to mention investigations on problems of recognition and identification in stochastic systems and also on problems of optimization of computations. These lines of investigations are actively developed at the Institute of Cybernetics over all the years of its existence, and we will still dwell on them in what follows.

Investigations in the field of economic–mathematical modeling were the object of constant attention of V. S. Mikhalevich. The constant need for the development of economic–mathematical models became especially appreciable at the new stage when Ukraine state began to master market economy mechanisms. Under his management, mathematical models of processes in a transition economy were elaborated, information technologies for decision-making support were developed, and software–hardware systems were created for practical realization of the models proposed. First of all, fundamentally new balance models were constructed and investigated that took into account the instability of prices that is organically inherent in a transition economy and that is accompanied by negative processes such as an increase in inflation and nonpayment and shortage of money in the sphere of finance.

To analyze processes in a transition economy, dynamic models in the form of systems of linear differential equations were considered that, as a rule, cannot be solvable in analytical form. Nevertheless, the application of methods of qualitative analysis, decomposition, definite analytical transformations, and numerical experiments made it possible to draw some conclusions on the solution of such equations. Based on the investigations pursued, inflationary situations such as inflationary demand crises and structural inflationary crises were analyzed. A comparative analysis testified to the adequacy of the proposed models to real-life processes.

Investigations of monetary and budget policy of the state were pursued. Models were proposed that allow one to establish an optimal money supply that, taking into account the dynamics of prices, does not create additional inflationary effects. The fact that this line of investigation is also being successfully developed at the Institute of Cybernetics at the present time is extremely significant, and a significant role in carrying out these investigations belongs to Mykhailo Volodymyrovych Mikhalevich who is the son of V. S. Mikhalevich and is now a corresponding member of NAS of Ukraine. We will consider the question of development of economic cybernetics in Chap. 4.

V. S. Mikhalevich was also agitated by ecological problems. This problematics became particularly topical after the Chernobyl catastrophe. As is well known, important works on Chernobyl themes were carried out under his guidance, and, in particular, a special software–hardware complex was created under the direction of A. O. Morozov in the Special Design Office (SDO) of the Ministry of Machine Building (MMS) at the Institute of Cybernetics on which current data on the state of pollution of the Chernobyl zone, Pripyat River, and Kyiv storage water basin were promptly processed and the process of diffusion of pollution was prognosticated. This complex became the main tool of analysis of situations and forecasting of consequences of the Chernobyl catastrophe. The attraction of many leading specialists to this problematics has led to the creation of a scientific line of investigations at the Institute of Cybernetics, which can be called computer technologies in ecology; they are being actively developed at the present time. We will dwell on some problems of development of this line of investigation in Chap. 3.

V. S. Mikhalevich devoted considerable attention to the development of the methodology of systems analysis and usage of its methods in solving complicated problems that arise in economy, management, design of complicated objects, and military science. Under his influence, this problematics was actively developed in a number of scientific departments of the Institute of Cybernetics and by specialists of other institutions, in particular, at the Taras Shevchenko State University of Kyiv and at the Kyiv Polytechnic Institute (KPI). On V. S. Mikhalevich's initiative, the Department of Mathematical Methods of System Analysis was created in KPI in 1988. Later on, taking into account the importance of the development of the theory and methods of systems analysis, the Institute of Applied System Analysis of the National Academy of Sciences of Ukraine and the Ministry of Education and Science of Ukraine was created on the basis of two scientific departments of the Institute of Cybernetics and the mentioned Department of KPI. This Institute was headed by M. Z. Zgurovsky, a well-known specialist in the field of system analysis and an academician of the National Academy of Sciences of Ukraine. We will consider some results of this line of investigation in the next chapter.

The essential and immediate role of V. S. Mikhalevich as the research supervisor of significant scientific–technical projects at branch and even intersectoral levels also manifested itself in numerous developments of application-specific systems that were carried out in the USSR at the Institute of Cybernetics and its engineering and design subdivisions. Let us give some examples that substantiate this.

Within the framework of execution of tasks assigned by the Ministry of Shipbuilding Industry of the USSR, a number of probabilistic problems of estimating the reliability and optimal redundancy of elements of technical systems were solved. Later on, but this time for the benefit of the Soviet Navy, a research and development (R&D) work (whose chief designer was P. I. Andon) of unique scope and volume was performed in which, on the scientific basis, a large number of operations research problems connected with assurance and optimization of reliability, maintenance, and logistical support of the Navy were solved. They included the practical problem of optimal planning of prospective and routine repairs of ships that was of the form of a partially block problem of Boolean linear programming with the number of unknowns close to 10^5.

One more large-scale R&D work at the branch-wise level in which V. S. Mikhalevich directly participated as its research supervisor was devoted to the solution of optimization problems for the Mission Control Center and the Cosmonaut Training Center of the USSR. In particular, within the framework of these R&D works, problems of optimal formation of scientific programs for space flights, optimization modeling of activity of cosmonauts in nonnominal situations, and also problems of optimal planning and management of the activity of space crews were first solved.

The breadth of interests of V. S. Mikhalevich and his acute feeling of new, urgent, and promising lines of investigations were largely formed under the influence of V. M. Glushkov. When the Computer Center was reorganized into the Institute of Cybernetics in 1962, Mikhalevich was appointed the deputy research director. He held this post until 1982. When V. M. Glushkov passed away, the entire burden of daily activities, obligations, uncompleted works, and advanced developments bore heavily on his shoulders. And he continued the cause begun by his outstanding predecessor with dignity.

One of the last projects of V. M. Glushkov was the development of a macroconveyor computer system, that is, a homemade sample of a supercomputer. At the end of the 1970s, the world electronics industry was faced with the problem of creation of a computer whose speed would be equal to billion operations per second and that would simultaneously perform computations on many processors. In the Soviet Union, several projects of this level were proposed. The macroconveyor computer complex (MCCC) developed at the Institute of Cybernetics was among them. Distinctive features of this project were original ideas of organization of parallel computations and internal mathematics in this computer that had no analogues in world practice. The mentioned circumstance turned out to be the major hindrance to the development and practical implementation rather than a considerable assistance in the realization of this project since, in the Soviet Union, unfortunately, preference was given to projects that copied Western patterns. This was reflected in the amount of financing of the macroconveyor complex. V. M. Glushkov had no time to overcome the difficulties connected with the completion of the development and commissioning of the MCCC. V. S. Mikhalevich had to do this. He mobilized the collective and used all available levers to successfully complete the works on the creation of the macroconveyor complex. In the second section of this chapter, we consider some works of this line of investigations that exerted a positive influence on the construction of computer technologies realized on supercomputer systems at a later date.

Deeply understanding the importance, capabilities, and prospects of development and usage of computer aids, as long ago as the 1960s, V. M. Glushkov suggested the idea of computerization of management processes at all levels beginning with a separate manufacturing enterprise and ending with an oblast and the entire state. But, at that time, this idea was considered to be fantastic since the available technical base did not make it possible to implement it in corpore. Nevertheless, definite investigations were pursued. Processes of creation of automated industrial management systems, organization of branch computer centers, their networking, and creation of unified data collection, processing, and transmission systems were investigated. All these investigations made it possible to provide for prospects of computerization of all types of human activities including manufacture, education, communication, business operations, housekeeping, etc.

Academician V. S. Mikhalevich

This idea was supported by the creation of at least the following two global projects: a united communication network of computer centers of the country and also a nationwide automated management system. V. S. Mikhalevich was the direct participant in the process of development of these projects.

It should be taken into account that these projects were developed under conditions of planned economy and a centralized state management system. Nevertheless, the situation essentially changed at the beginning of the 1990s. The onrushing advent of inexpensive personal computers with a rather high-performance and high-speed communication facilities made the computerization of the society not only possible but also necessary. And, at the Institute of

Cybernetics headed by Mikhalevich, works on the creation of a society informatization program oriented toward new scientific and technical capabilities and economic–political conditions of Ukraine as an independent state were again resumed. This program was developed through the efforts of specialists of the Institute of Cybernetics, other establishments of the National Academy of Sciences of Ukraine, branch institutes, and representatives of state organs.

It should be noted that this work was supported by the state government. V. S. Mikhalevich was appointed the adviser to the President of Ukraine on informatization. The work was regarded with favor and was supported by the Cabinet of Ministers and the Supreme Rada (Parliament) of Ukraine. To practically realize the National Society Informatization Program, a special state organ called the Agency for Informatization was established under the President of Ukraine. Later on, the Supreme Rada adopted the Informatization Program and the Law on the National Program of Informatization of Ukraine. In 1998, the Supreme Rada of Ukraine created its Advisory Council for Informatization. It would seem that all conditions for the practical realization of the National Program of Informatization of Ukraine were created. Nevertheless, the main resources, that is, necessary material resources, were absent. Under the current complicated economic conditions, one can understand this fact. Nevertheless, it is impossible to understand and accept the fact that even insignificant resources allocated to the informatization program are not always efficiently used and do not conform with adopted normative documents.

V. S. Mikhalevich clearly realized the needs of the state and mobilized the collective for the execution of urgent tasks and solution of advanced problems. He was a scholar of authority not only in our country; he regularly participated in international scientific forums, closely cooperated with the International Institute for Applied System Analysis, headed the National Committee on Systems Analysis, and was a member of the European Association for Problems of Risk. He was well known in and respected by the international scientific community. He also was an active public figure and was elected a deputy to the Supreme Rada of the UkrSSR.

Eminence in science achieved by Mikhalevich was awarded State Premiums of Ukraine and Soviet Union and the N. M. Krylov, V. M. Glushkov, and S. A. Lebedev Prizes of the National Academy of Sciences of Ukraine.

He was a member of the Presidium of NAS of Ukraine and the academician-secretary of the Department of Informatics organized by him with the active support of B. E. Paton, the President of NAS of Ukraine. The Presidium of NAS of Ukraine instituted the prize named after V. S. Mikhalevich. In 2000, this prize was first awarded to his disciples Academicians Yu. M. Ermoliev, I. N. Kovalenko, and N. Z. Shor.

Mikhalevich devoted himself to the zealous service to science and loved the Institute of Cybernetics to which he gave a significant part of his life.

At the present time, probably, not all the achievements of V. Mikhalevich in science are thoroughly assessed. They will obviously be comprehended in future.

1.2 First Supercomputer Complexes and Parallel Computations

In the Institute of Cybernetics, investigations in the field of new computer architectures and supercomputer complexes began way back in the 1970s. The report of V. M. Glushkov together with M. V. Ignatiev, M. A. Myasnikov, and V. A. Torgashev at the IFIP Congress in Stockholm in 1974 [223] gave the first impetus to these investigations. This report was devoted to the revision of Neumann's principles and to the presentation of a new architecture of computers, which was called recursive. In contrast to the von Neumann sequential architecture, they proposed the principle of recursive construction of a computer at all levels of its structure from the uppermost level of source programming languages to communication systems that provide interconnections between processor elements. The consistent use of this principle promised the possibility of unbounded parallelization of programs and construction of supercomputers on the basis of new architectural principles.

The authors of the report strived to realize their ideas as soon as possible, and they immediately began to search for a practical embodiment of the idea of a recursive computer. But the technology of that time was still not ready for this work. They were forced to search for compromises. Such a compromise was found in 1978. V. M. Glushkov proposed a new principle of parallel computations and called it the *principle of macroconveyor*. In contrast to a microconveyor that was already used in vector supercomputers and was oriented toward the parallelization of inner loops of programs, a macroconveyor was destined for the parallelization of outer cycles. It was expected that, for problems of superhigh complexity, this principle would make it possible to obtain a linear increase in computer speed with increasing the number of processor elements. To support macroconveyor computations, one should only connect a rather large number of conventional Neumann processors with the help of a universal communication system providing "all-to-all communications." It should be noted that a realization of even this very simple architecture was connected with the necessity of solution of several technical problems since fast communication systems that provide communication in modern clusters did not exist yet. Mathematical system software and methods of macroconveyor computations for concrete object domains also had to be developed.

Beginning in 1979, the collective of developers of the future macroconveyor system was being formed at the Institute of Cybernetics. The departments of V. M. Glushkov, V. S. Mikhalevich, I. N. Molchanov, I. V. Sergienko, E. L. Yushchenko, and others were involved in this project. The preliminary design was developed in the Special Design Bureau (SDB) of Mathematical Machines and Systems. S. B. Pogrebinskii was appointed to the position of the design manager of the project. The collective of engineers headed by him already had an experience in the development of MIR series computers.

Simultaneously with the formation of the collective of developers of the macroconveyor, V. M. Glushkov persistently and successfully got in contact with industry. He managed to attract the Ministry of Radio Industry of the USSR and the Penza computer factory to the execution of the project. The project financed through the Research-and-Development Center of Electronic Computer Engineering that, in particular, was responsible for the production of the unified system of electronic computers (ES EVM), and the project itself was included in the program of development of ES EVM.

V. M. Glushkov's hope to see the functioning of the macroconveyor did not materialize. The execution of the project was headed by V. S. Mikhalevich, the new director of the Institute of Cybernetics, who triumphantly accomplished it. The project was completed in 1987 and was passed to series manufacturing. The disintegration of the Soviet Union stopped the production of the first computing system of macroconveyor type in the world. But the experience obtained during the development and usage of the macroconveyor was not in vain and is widely used in modern technologies of solution of complicated problems but already on modern cluster and grid systems.

The problems that had to be solved by V. S. Mikhalevich were not simple from the very beginning. In developing the technology of macroconveyor computations, only first steps (mostly at the theoretical level) were made. They should be tested on authentic applied problems. An original system software whose analogues were absent among well-known systems had to be developed for the macroconveyor. Technical problems had to be solved. Exceptional efforts had to be made to communicate with the external environment, to support ties to industry, to search for future users of the macroconveyor, and to competitively struggle for a place among other projects of supercomputers that were developed in the Soviet Union at that time.

Operational meeting. Yu. V. Kapitonova, V. S. Mikhalevich, and I. V. Sergienko

The first task consisted of supporting and strengthening close interaction between all collectives involved in the work on the project. Regular scientific seminars and organizational meetings promoted the solution of complex scientific problems and current questions connected with the timely completion of works.

The development and manufacturing management of developmental prototypes of the technical equipment for the macroconveyor was performed by A. O. Morozov, the director of the SDB; S. B. Pogrebinskii (design manager); A. G. Kukharchuk (communication system); V. P. Klimenko (internal software and control processors); and A. I. Tolstun (preparing the design documentation for series manufacturing).

I. N. Molchanov was responsible for the external software and preparation of applied problems for the execution on the macroconveyor. Yu. V. Kapitonova and A. A. Letichevskii were responsible for the development of the system software and computer-aided design. At that time, the author worked on the statement and investigation of some important discrete optimization problems with a view to solving them on the macroconveyor.

V. S. Mihalevich personally participated in investigations on the development of the methodology and technology of computations on multiprocessor systems of macroconveyor type. He is a coauthor of two important articles [93, 94]. They continued and deepened ideas of V. M. Glushkov on macroconveyor computations that had already passed their first practical tests. At that time, the following idea prevailed that was stated by Marvin Minsky in the 1970s and was supported by M. Flynn in his well-known classification of computer control systems: Increasing the number of processors, one can obtain only some logarithmic increase in the speed of multiprocessor computer complexes. Dispelling this myth, a new classification of computer architectures that was more exact than the Flynn classification was presented in [93]. It showed that, in MIMD (multiple instruction stream/multiple data stream) systems with a universal communication system and distributed memory, it is possible to obtain a linear increase in speed owing to the macroconveyor organization of computations for problems that require the execution of a rather large number of operations. The mentioned article also presented main principles of construction of the parallel programming language MAYAK and an operational system supporting macroconveyor computations.

In [93], fundamental schemes of organization of parallel computations are analyzed; such schemes are called patterns for parallel programming at present and are proposed as a result of analysis of many parallel algorithms used for solving applied problems.

The simplest form of organization of macroconveyor computations is a static macroconveyor. It consists of a definite number of components each of which performs a cycle of computations and data exchanges with other components depending on the topology of information connections between them. Different kinds of static macroconveyors are realized under different conditions of organization of exchanges. First of all, it is the organization of synchronous and asynchronous data transmission. If all components begin to operate immediately after calling the static macroconveyor, we have an iterative version. If each component waits for information incoming from an adjacent component before beginning its computation, then we obtain a wave macroconveyor. In each cycle, iterative and wave macroconveyors are switched only once from computations to data transmission and reception. This makes it possible to rather exactly estimate the efficiency of parallel computations with the help of the mentioned schemes of their organization.

A mixed macroconveyor is characterized by the capability to repeatedly change its modes of exchange and computations in arbitrary order. It is not so easy to estimate the efficiency of a mixed macroconveyor, and, moreover, it can assume dead states whose disclosure requires the use of special methods of analysis.

A static macroconveyor allows for an efficient realization of a considerable number of computational methods including methods of linear algebra, mathematical physics, optimization, etc. Schemes of a dynamic macroconveyor with varying connections, components, and exchange modes can be more flexible. Some special schemes of a dynamic macroconveyor can be investigated in more detail in comparison with the general case. A modeling macroconveyor, dynamic parallelization, and some others are such schemes. They form multilevel and recursive networks that already approach structures of a recursive computer and brain-like structures about which V. M. Glushkov dreamt. These schemes can be used for modeling complicated system structures, artificial intelligence systems, etc.

To support macroconveyor computations, the language and system of parallel programming MAYAK, a distributed operational system that supported the realization of main constructions of this language, and the corresponding translators and programs of internal software were developed.

The first product of the project was the macroconveyor computer complex (ES MCC) developed as a specially designed processor ES 2701 destined for the increase in the performance of basic models of ES EVM. The MCC included processors of the following types:

- Arithmetic processors
- Control processors
- Peripheral processor
- Communication processor

The ES MCC included one peripheral processor, which was one of top-of-the-line models of ES EVM. Any pair of processors could communicate among themselves through a communication network controlled by the communication processor. The peripheral processor supported the interaction between terminals and external devices.

The system of commands of an arithmetic processor was close to the ES EVM system of commands extended by vector operations. An arithmetic processor has internal memory sufficient for the organization of efficient macroconveyor computations.

A control processor ES 2780 provided hardware support for a high-level language and was destined for the realization of functions of control over groups of arithmetic processors. Control processors realized components of the distributed operational system. They supported program loading, synchronization of exchanges, and elimination of malfunctions and errors. Control processors provided the allocation of processors among tasks and formation of macroconveyor networks of various types. Control processors could also be used as intelligent terminals.

The system of source programming languages included the parallel programming language MAYAK and traditional (at that time) languages such as Fortran, PL/1, and COBOL. A language of directives was the main tool of access to the system and allowed one to work in batch and interactive modes. In particular, the language of directives performed all functions of a task-control language.

The operational system provided the operation of technical facilities and their interaction with programs and data that formed the information environment of the ES MCC. The environment included a data formation system that could create and form the information environment irrespective of concrete problems being solved.

The programming system provided the translation of source languages and interaction between program modules and data modules of the ES MCC information environment. During the translation of parallel programs, all necessary modules of the operational system and modules of interaction with the information environment were created. The extensions RPL/1 and PCOBOL were created for the languages PL/1 and COBOL to support parallel programming tools in these languages [23].

The system of support tools of the MCC was used as the base for the development and extension of the system-wide and applied software of the MCC and for the creation of applied programs and specialized software systems.

The developed ES MCC (of the model ES 2701) with the macroconveyor organization of computations and a processor of the model ES 2680 for the interpretation of high-level languages in the capacity of an intelligent terminal and an extender of the computational environment was approved by the State Commission in 1984.

The next step consisted of developing and productionizing the ES MCC (of the model ES 1766) with macroconveyor organization of computations. This task was completed in 1984 under the direction of V. S. Mikhalevich. During the delivery of the MCC to the State Commission, the record (at that time) performance 532 MFLOPS for 48 processors and the possibility of a linear increase in the performance with increasing resources were fixed. In particular, it was shown that, for the designed number of arithmetic processors, the MCC performance could be increased up to 2 GFLOPS. Thus, the ES MCC of the model ES 1766 was the first supercomputer complex in the Soviet Union (the supercomputer "Elbrus" of the Institute of Precision Mechanics and Computer Engineering of the Academy of Sciences (AS) of the USSR was approved 6 months later).

To solve scientific and technical problems on the ES MCC, algorithms of parallel computations and programs for the solution of problems of computational mathematics of the following classes were developed in the Soviet Union for the first time: high-dimensional systems of linear algebraic equations (solved by direct and iterative methods), partial and complete eigenvalue problems, systems of nonlinear equations, Cauchy problems for systems of conventional differential equations, and approximations of functions of one and several variables.

In parallel methods of triangulation and tridiagonalization of matrices, block-cyclic data decomposition and processing algorithms were proposed for the first time, which allowed one to create balanced parallel algorithms that solved systems of linear algebraic equations and the algebraic eigenvalue problem by direct methods.

The MCC was supported by specially developed iterative methods for the solution of difference linear and quasilinear elliptic equations, difference methods for the solution of evolutionary problems (difference analogues of equations of parabolic and hyperbolic type), and parallel algorithms and programs for the numerical solution of integral equations.

For optimization problems, parallel algorithms and programs were created that solved the quadratic assignment problem, cluster analysis problems, high-dimensional production–transportation planning problems, and nonsmooth optimization problems. Parallel algorithms for solving linear and linear Boolean programming problems were created. A general scheme of parallel computations was developed for algorithms of decomposition of optimization problems. A decomposition algorithm was created for solving linear programming problems of partially block form. Parallel branch-and-bound algorithms were developed for the solution of high-dimensional Boolean programming problems.

Moreover, numerical methods were developed to simultaneously solve some problems of flow past a given body, including problems of transonic flow past aerodynamic bodies and viscous incompressible fluid flow past an elliptic cylinder at an angle of attack.

A number of highly complicated applied problems were solved on the ES MCC, namely, problems of research into the strength of an airframe as a whole (they were considered for the benefit of the M. E. Zhukovskii Military Aviation Engineering Academy), numerical modeling of nuclear explosions (using a program of the All-Union Scientific Research Institute of Experimental Physics), investigation of ocean–atmosphere interactions (using a program of the Department of Computational Mathematics of AS of the USSR headed by G. I. Marchuk). The obtained results illustrated a high efficiency of the MCC and a good correlation between the results of numerical and natural experiments.

V. S. Deineka constructed a mathematical model and developed a computational scheme for calculation of deflected mode of an airframe; an analysis of the realization of this model in the mode of parallel processing of fully connected mass data promoted the implementation of new important functions created in this connection and embodied in the model ES 1766 of ES EVM. With the help of this computational scheme, the formation of the deflected mode of the airframe of the IL-86 aircraft as a whole was analyzed for the first time under operating conditions on the ES MCC. The energy functional for this body Ω (a plate of piecewise-constant thickness) was obtained in the form

$$\Phi(U) = \frac{1}{2}\int_{\Omega}\left(\lambda\sum_{i=1}^{2}\left(\frac{\partial u_i}{\partial x_i}\right)^2 + 2\lambda v\frac{\partial u_i}{\partial x_1}\frac{\partial u_2}{\partial x_2} + \frac{\mu}{2}\sum_{i,j=1 i\neq j}^{3}\left(\frac{\partial u_j}{\partial x_i} + \frac{\partial u_i}{\partial x_j}\right)^2\right)dx$$

$$-\sum_{i=1}^{4}Q_i e_3(d_i),$$

where λ, $\mu = f(E,\ v)$, E, and v are, respectively, the Young modulus and Poisson ratio; u_i is the projection of the displacement vector U onto the ith axis of the Cartesian coordinate system; and Q_i is the magnitude of the force applied to the body $\bar{\Omega}$ at the point d_i.

V. S. Mikhalevich gave a great attention to the establishment of ties with potential users of the ES MCC. Together with I. N. Molchanov, they visited many scientific and branch enterprises to familiarize the community with new supercomputer technologies and investigated needs for supercomputer computations in various application domains.

A special role was played by contacts with the M. E. Zhukovskii Military Aviation Engineering Academy. Seminars on the investigation of methods of macroconveyor computations were organized together with specialists of this institution. With the support of the Academy, two scientific–methodical monographs [165, 180] were published, which were the first monographs on parallel computations in the Soviet Union. Practically all the developers of the system and applied software of the MCC participated (as coauthors) in preparing these monographs.

At that time, several projects of supercomputers were developed in the Soviet Union. Among them, the most well-known projects were "Elbrus" (at the Institute of Precision Mechanics and Computer Engineering of AS of the USSR under the guidance of the chief designer V. S. Burtsev), the family of computers PS (at the Institute of Control Sciences of AS of the USSR under the direction of I. V. Prangishvili), and MARS (at the Siberian Branch of AS of the USSR under the guidance of G. I. Marchuk).

The Elbrus project enjoyed the greatest support of the government. It was financed by the Ministry of Electronic Industry (MEI) in which the most advanced element base was developed. The amount of financing of the Elbrus project was by several orders of magnitude greater than that of the macroconveyor. Therefore, to maintain the existence of the ES MCC project, V. S. Mikhalevich, the project manager, had to be permanently active in all scientific and organizational actions, beginning with scientific conferences and ending with sessions at the level of the Academy of Sciences of the USSR, ministries, and executive departments. V. S. Mikhalevich and scientists of the Institute of Cybernetics resolved these questions with great responsibility.

Time has shown that the line of development of distributed supercomputers that was initiated by V. M. Glushkov has turned out to be most competitive. Modern cluster systems whose first representative was the macroconveyor of the 1980s form the main direction in the construction of supercomputers at the present time.

Scientific research into technologies of parallel computations and their application based on a new technical basis continued (and is being continued now) after the macroconveyor became a part of history. The development staff (I. V. Sergienko, N. Z. Shor, A. A. Letichevskii, Yu. V. Kapitonova, I. N. Molchanov, V. A. Trubin, and V. P. Klimenko) headed by V. S. Mikhalevich was awarded the State Prize of Ukraine for the cycle of works "Mathematical methods and software tools for parallelization and solution of problems on distributed multiprocessor computers" in 1993.

A new impetus to the development of technologies for solving problems requiring high-performance computing turned out to be the widespread use of cluster systems. At present, the Institute of Cybernetics performs extended studies in this line of investigation. Now, the V. M. Glushkov Institute of Cybernetics not only has SCIT-3, which is one of the most powerful cluster systems in the Commonwealth of Independent States (CIS) (the chief designer of the system is V. M. Koval) but also possesses a few tens of technologies that make it possible to very efficiently solve most complicated problems of economy, ecology, information security, environment protection, space research, inquiries into regularities of biological processes, etc.

Supercomputers of the SCIT family are combined into a supercomputer complex whose performance is of the order of 7 TFLOPS. It serves computational needs of the scientific community of NAS of Ukraine and universities through the Internet and is a component of an international grid system. This allows one to solve fundamentally new problems of transcomputational complexity in the fields of nuclear research and artificial intelligence, creation of new materials, investigation of the human genome, meteorological forecasts and prediction of natural disasters, nanotechnology, and other fields.

At the present time, the cluster configuration is composed of 125 nodes, 250 processors, 704 nuclei, and 1.4 TB of main memory. The theoretical performance of the cluster reaches 7.4 TFLOPS, and its actual performance evaluated with the help of the LINPACK test reaches about 6 TFLOPS.

In 2009, SCIT supercomputer powers were completely used by scientists of institutes of NAS of Ukraine and higher educational institutions. When used skillfully, they allow one to more efficiently solve problems of prevention of ecological catastrophes connected with the influence of human economic activity on the environment, problems of processing and interpretation of geophysical data to prospect for oil and gas, weather modeling and forecast, and also consequences of harmful emissions to the atmosphere. The Supercomputer Computation Center solves problems of processing space photos obtained by the satellite "Meteostat" to forecast cloudiness and origination of extreme meteorological situations whose tracing is important for planning air traffic and aerospace photography. Calculations performed by the Institute of Space Research of NAS of Ukraine on SCIT-3 make it possible to efficiently use data obtained by the satellite "SICH-IM" and to assist in providing the efficient use of the satellite "SICH-2" and also in

Table 1.1 Technical characteristics of intelligent computers

	Name			
	Inparcom 32	Inparcom 64	Inparcom 128	Inparcom 256
	Processors			
	Xeon quad-core	Xeon quad-core	Xeon quad-core	Xeon quad-core
Number of nodes/ processors/cores	4/8/32	8/16/64	16/32/128	16/64/256
Peak performance, GFLOPS	200–380	400–770	800–1,500	2,385
LINPACK performance, GFLOPS	150–290	300–580	600–1,130	1,855
Main memory, Gb	64	128	256	512
Disk memory, Tb	1	2	4	8
Disk repository, Tb	—	≥ 1	≥ 2	≥ 3

optimally choosing orbits and planning the functioning of Earth remote sensing satellites for the purpose of land use, protection of large forests, and monitoring of the development of elemental and technogenic processes.

Taking into account a high efficiency of macroconveyor data processing, the main principles of construction of the model ES 1766 were used in creating (together with the State Research-and-Production Enterprise "Elektronmash") intelligent workstations INPARCOM, which occupy the niche between personal and supercomputers. This development was headed by V. I. Mova, the general director of the SRPE "Elektronmash," and Professor I. N. Molchanov. The development of computational and system software and technical means was carried out by the departments of O. M. Khimich and O. L. Perevozchikova at the Institute of Cybernetics and the Research-and-Production Enterprise "Poisk" headed by V. A. Stryuchenok. This work has been carrying out since 2005 up to date.

INPARCOM is a family of intelligent workstations whose structure and architecture and also operational environment support intelligent software. The operational environment allows a user to optionally install operational systems Linux or Windows. The interprocessor interaction is realized with the help of MPI. The programming languages C, C++, and Fortran are used. Intelligent software realizes the function of automated adaptive adjustment of the algorithm, program, and topology of the intelligent computer to the properties of the computer model of a problem that are investigated during the obtaining of a computer solution [179].

The technical characteristics of intelligent computers are presented in the following table (Table 1.1).

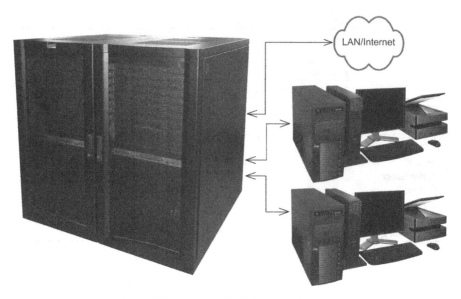

Intelligent workstation INPARCOM 256

These workstations were used to solve problems of flow over the airframe of the AN-148 aircraft (the O. K. Antonov Aeronautical Scientific/Technical Complex), some problems of construction of gas turbines (the state enterprise "Ivchenko Progress"), and some problems of electric welding (the E. O. Paton Electric Welding Institute of NAS of Ukraine) and to perform the strength calculation of an office center in Moscow, which showed a correlation between numerical and full-scale experiments and efficient functioning of INPARCOM intelligent workstations.

1.3   Solid Foundation for Computer Technologies

A grain casted into the ground spires, sprouts, and yields a good harvest later on. Knowledge obtained does not remain only on pages of monographs and in texts of scientific articles but is used for the benefit of the people and for human development.

Excellent examples of the aforesaid are mathematical theories and optimization methods. From the outset, the need for them arose in all cases when it was necessary to choose the best solution among many alternate solutions to a complicated problem. The activity of specialists in optimization met with approval in society and was highly appreciated by bodies of state authorities. Cyberneticists, authors of most successful works, were repeatedly awarded State Prizes of Ukraine and the USSR and other honorable distinctions.

Since the number of complicated problems permanently increases, their solution requires the development of corresponding new methods. The school of optimization is extended and increases its influence on a variety of spheres of human activity. Works of disciples and followers of V. S. Mikhalevich are still topical at the present time, and their results are intensively embedded into routine practice. The award of the State Prize of Ukraine in the field of science and engineering to the composite authors of the cycle of works "Development of new mathematical models, methods, and information technologies for solving optimization problems of information processing and protection" in 2009 testifies, in particular, to this fact. This group consists of leading scientists of the V. M. Glushkov Institute of Cybernetics of the National Academy of Sciences of Ukraine, Taras Shevchenko National University of Kyiv, National Technical University of Ukraine "KPI," and educational institutions of Poltava and Kharkov.

The authors of the mentioned cycle of works are well-known specialists, namely, corresponding members of NAS of Ukraine V. Zadiraka and S. Lyashko and doctors of sciences L. Gulyanitskii, G. Donets, O. Yemets, P. Knopov, A. Levitskaya, O. Litvin, D. Nomirovskii, and A. Pavlov; they are representatives of the Ukrainian school of informatics, which is well known in the scientific world.

Taking into account the fact that we discuss an important and very promising trend in embedding modern optimization methods, let us consider it in more detail. Results of this line of investigation were repeatedly reported by authors at international scientific forums and were issued by leading domestic and foreign publishers.

One of the most developed topics of this line of investigation is the theory of decision making. It is obvious that a successful development of a society is impossible without daily competent decision making at all levels of its functioning. This is true for global problems at the state administration level, problems of market economy, and local problems of ecology and concrete manners of manufacture, problems of public life, problems of bank and financial activity, etc.

Decisions are made on the basis of previous experience and available information and with allowance for uncertainty and risk factors connected with the inaccuracy of data, absence of precedents, limited prediction capabilities, and uncertain nature of the phenomenon itself being investigated. A predestination of science is the creation of tools for quantitative estimation of uncertainty and risk, means for calculation of the estimates obtained, and risk control methods.

Traditional scientific approaches are based on real-life observations and experiments. There is no sufficient number of observations for problems under uncertainty, and experiments and investigations performed by the trial-and-error method can be very expensive, dangerous, or not possible at all. These problems require modeling uncertainty using system information composed of available reliable data of previous observations, results of possible experiments, and scientific facts. However, an analysis performed in this way always has a limited accuracy, and the role of science in solving new problems increasingly evolves from traditional deterministic predictions to the production of strategies that take into account uncertainty and risk.

The composite authors obtained profound fundamental results that underlie a newly developed mathematical apparatus for the analysis of complicated systems under risk and uncertainty. This apparatus generalizes and improves widely well-known procedures of optimization and decision making, includes new methods of estimation under conditions of insufficient information, and proposes computational algorithms and software tools for solving supercomplicated problems of stochastic optimization.

A significant part of the results was obtained in cooperation with the International Institute of Applied Systems Analysis (IIASA, Vienna) whose projects were performed with the direct participation of the composite authors. This appreciably supports the international prestige of Ukraine that cooperates with IIASA over 30 years.

Based on a system combination of approaches and methods of decision-making theory, combinatorial analysis, and discrete and stochastic optimization, specialists of the V. M. Glushkov Institute of Cybernetics created a computer technology that supports making crucial decisions and was applied to strategic planning and management, design and development of systems of production and commercial activity, and also in analyzing, modeling, and predicting consequences of actions in solving some economic and sociopolitical problem or other.

In particular, for successful operation of atomic power plants under conditions of insufficient information, new algorithms of estimation of reliability parameters are proposed whose accuracy considerably exceeds that of previously used methods and algorithms. A number of projects were proposed and carried out that support making crucial decisions and are directed toward the transformation of the economy of Ukraine into a market economy, control over complicated hi-tech systems, systems for computer-aided production management in small-scale manufacturing businesses, and modeling and medium-term forecasting of the dynamics of changing main macroeconomic indicators of the economy of Ukraine.

Decision-making support models for export–import policy, formation of investment priorities, search for economic equilibria, estimation of the behavior of a foreign investor under risk and uncertainty, and economic dynamics of the labor market and also models of financial mathematics are developed and implemented. Some of these projects were carried out under contracts with the Ministry of Economy and European Integration of Ukraine, Ministry of Extreme Situations of Ukraine, and Ministry of Education and Science of Ukraine.

The results of forecasting the real GDP of Ukraine showed that the developed model has a high confidence level in comparing forecasted and actual results of the development of the economy of Ukraine and that this level surpasses even the well-known short-term (annual) forecasts made by specialists of such well-known organizations as the International Monetary Fund (IMF), World Bank, European Bank for Reconstruction and Development, etc. Thereby, the possibility of adaptation of components and means of the created information technology to new classes of problems has been practically confirmed. The available actual data testify that the developed mathematical models and created technology of system analysis make it possible to obtain a fundamentally higher confidence level of results. Therefore, the developed methodology can become the base for further investigations of the economy of Ukraine.

An important result of the cycle of works is also the development of theoretical, methodological, and algorithmic foundations for the passage from classical to neoclassic computing methods by parallelization of algorithms of computations with succeeding aggregation of results. In particular, for concrete practical problems concerned with the investigation of formulas of chemical compounds, formulas or circuits implementing Boolean functions, digital automata, and polytopes, a special theory of numerical graphs is developed that uses properties of objects being investigated such as symmetry, periodicity, or similarity of some fragments. The theory created provides the possibility of constructing efficient algorithms for the solution of many complicated problems. Its application to the investigation of mathematical models of parallel computing systems turned out to be particularly successful.

The intensive development of the theory of numerical graphs in the V. M. Glushkov Institute of Cybernetics dates back to the 1980s. And it has become an independent subdiscipline of the general theory of graphs.

In the theory of recognition of complicated images, various practical problems of their analysis arise that do not fit into general frameworks for classification. A mathematical theory of construction of discrete images with the help of color patterns was developed for the first time. This theory can be used, in particular, in creating some cryptographic systems. Efficient approaches to the problem of computer recognition of visual images are proposed. Hundreds of scientific teams and companies deal with this problem; nevertheless, it remains far from its final solution. Within the framework of the cycle of works being considered, a collection of means is developed that form a model of an image. Based on this model, recognition algorithms are formed on the basis of not only intuition or experience of a researcher but also on the basis of the results of solution of concrete mathematical problems.

A wide class of problems of Euclidean combinatorial optimization is investigated to which many optimization problems in economic, transport, agricultural, medical, scientific, military, humanitarian, social, educational, and other complex systems are reduced. The apparatus developed by the authors for modeling and solving problems of the mentioned class is a tool that makes it possible to exactly and (in the overwhelming majority of cases) efficiently solve important problems of optimization modeling in a new way, with best results, and in various fields of science and practice.

At each stage of development of computational capabilities and methods for solution of applied problems, mathematical problems of superlarge dimensionality arise (such as control over rapidly varying processes, radar data processing, reactor diagnostics, determination of the maximum permissible level of accumulation of methane in coal mines, etc.) that should be solved in real time. To overcome the barrier of transcomputational complexity, the use of various reserves of optimization of computations with the help of development of methods optimal with respect to their accuracy and speed and with the help of creation of powerful computers is always topical. It is established that, for some classes of problems (digital signal processing, pattern recognition, etc.), the effect of using optimal methods and also models of parallel computations is comparable with the effect of using a more modern element base of computers.

The process of informatization of Ukraine has attained the stage at which the problem of development and embedding of supercomputer technologies urgently arises. Every effort of the recently created State Committee for Informatization of Ukraine and Cybernetic Center of the National Academy of Sciences of Ukraine (this center is prescribed to be the chief scientific institute for solving these questions in the country by the corresponding Decree of the President of Ukraine as long ago as 1992) must be applied to the implementation of this program.

An urgent problem arose in the country in due time; it consisted of the improvement of tools that make it possible to develop and construct systems of cryptographic protection of information and conform to modern world standards. Investigations in this topical direction are intensively pursued in the V. M. Glushkov Institute of Cybernetics. A number of important problems have been considered that arise in analyzing the resistance of systems of information protection in state and private sectors of the economy. The successful use of the apparatus of the algebraic theory of groups and combinatorial analysis made it possible to develop and apply the corresponding mathematical apparatus and also to propose efficient methods for the solution of such problems, which is a constructive contribution to the theory and practice of modern cryptography. This applies to systems of symmetric and asymmetric cryptography and computer steganography. Corresponding works also reflect the obtained and implemented results concerning the increase in the performance of systems of public-key cryptography (e.g., in attaching electronic digital signatures) and the development of new resistant means for secure message transfer and intellectual property protection.

The composite authors mentioned above also developed new methods optimal with respect to accuracy and operation speed in solving problems of digital signal processing, in particular, problems of spectral and correlation analysis of random processes, computation of discrete transformations, and problems of revealing the hidden periodicity. The obtained theoretical results and corresponding computer technologies are approved and used in solving problems of fast processing of a flight experiment performed under a contract with the Ministry of Defense of Ukraine and in educational processes of higher schools of Ukraine.

In a number of works devoted to the processing of multidimensional signals, new methods of approximation theory for functions of many variables are proposed. Based on them, new exact methods for solving problems of computer tomography were created that, in particular, allow one to improve the diagnostics of diseases with the help of information obtained from medical computer tomographs. The authors obtained patents that make it possible to use the mentioned methods at the customs for nondestructive testing of objects.

Other obtained theoretical results were also practically used in various establishments such as the Ministry of Economy of Ukraine, Research-and-Production Association "Impulse," State Scientific-Technical Center of Nuclear and Radiation Safety, Higher Command-Engineering Military School of Rocket Forces (Kharkiv),

Association "Motor-Sich" (Zaporizhia), "Zorya-Mashproekt" enterprise and VATT "Damen Shipyards Okean" shipbuilding enterprise (Mykolaiv), Poltava University of Consumer Cooperation of Ukraine, Research Institute of Economy of the Ministry of Finance of Ukraine, "Energoryzyk" Ltd, etc.

This brief review is quite sufficient to draw the conclusion that Ukrainian scientists made a considerable contribution to world fundamental and applied sciences, which opens qualitatively new vistas for solving supercomplicated problems of various natures.

Chapter 2
Optimization Methods and Their Efficient Use

Abstract The main scientific interests of V. S. Mikhalevich were connected with investigations in optimization theory and system analysis and with the development of newest computer technologies and computer complexes, creation of scientific bases for the solution of various problems of informatization in the fields of economy, medicine, and biology, design of complicated processes and objects, environmental research, and control over important objects. An analysis of the results of investigations along these directions is the subject matter of this chapter. Mikhalevich's great interest in the development of system analysis methods is shown. They began to be developed with his active participation at the Institute of Cybernetics of the National Academy of Sciences (NAS) of Ukraine, International Institute for Applied Systems Analysis (Austria), National Technical University of Ukraine "Kyiv Polytechnic Institute" (NTUU "KPI"), Taras Shevchenko National University of Kyiv, and other scientific and educational establishments of Ukraine. The problematics of this important line of investigations became especially topical in recent decades in connection with investigations of complicated processes of international cooperation in economy, problems of prediction and prevision of possible development of a society under conditions of environmental contaminations, and investigation of other complicated processes from the position of system analysis. In investigating such processes, an important role is played by the Institutes of the Cybernetic Center of NAS of Ukraine and also by the above-mentioned International Institute for Applied Systems Analysis whose creation was directly connected with V. S. Mikhalevich. At present, many scientists from Ukraine and many other countries cooperate with this institute. This chapter elucidates the history of creation of this institute, principles of its work, and the method of financing of important projects. Some important investigations of Ukrainian scientists are also considered that were performed during recent 15–20 years and that consisted of the development of new methods of stochastic and discrete optimization and the further development of the scheme of the method of sequential analysis of variants.

I.V. Sergienko, *Methods of Optimization and Systems Analysis for Problems of Transcomputational Complexity*, Springer Optimization and Its Applications 72, DOI 10.1007/978-1-4614-4211-0_2, © Springer Science+Business Media New York 2012

2.1   General Algorithm of Sequential Analysis of Variants and Its Application

Since the late twentieth century, achievements in exact sciences have been applied more and more widely to support managerial (economic, design and engineering, or technological) decisions. They help experts to understand the deep interrelation of events and to make the analysis and prediction more reliable. This allows public authorities, design and engineering organizations, and individual economic entities to find efficient decisions in complicated and ambiguous situations, which are so numerous nowadays. These applications became most popular in the 1940s. That was promoted by the rampant development of applied mathematics and computer science; nevertheless, the successful use of such tools required developing formal (including mathematical) decision-support methods, among which optimization methods occupy a highly important place. Since the late 1950s, the Institute of Cybernetics has played a leading role in the development of such methods. In many respects, credit for this goes to V. S. Mikhalevich, one of the most prominent scientists of the institute, an academician, and the director of the institute (from 1982 to 1994) who headed the studies in this field.

Mikhalevich's first scientific studies [24, 25, 100] were concerned with empirical distribution functions. The studies he carried out later, in Moscow, dealt with sequential procedures of statistical analysis and decision making [98, 103], which was a new field initiated by A. Wald [13], who proposed sequential models for making statistical decisions, taking the cost of experiment into account. Mikhalevich applied Wald's ideas to practically important acceptance sampling problems in Bayesian formulation where the probability distribution of rejects is known a priori. Different results of experiments lead to different a posteriori distributions, and it is possible to compare, at each step, the average risk of continuing experiments considering their cost and the risk of making a certain decision at the same step. The assessment of the average risk involves complex recurrent relations similar to the Bellman–Isaacs dynamic programming equation [11] (note that there was hardly any relevant American scientific literature at that time). Mikhalevich succeeded in using the concept of "reduced" (in the number of steps) procedures to derive pioneering results in acceptance sampling and reported these results in the paper [103], which actually outlined the key results of his Ph.D. thesis [104]. At the Kyiv State University, he wrote the theoretically interesting paper [99] where he treated continuous-time stopping problems at a high mathematical level. These problems were later developed by A. N. Shiryaev [182] and his disciples.

At the university, Mikhalevich continued his investigations into acceptance sampling problems. Since determining the decision-making limits in a significant number of sequential steps required labor-intensive calculations, he took an interest in the capabilities of the MESM computer, which had recently been launched in Feofaniya (a Kyiv suburb). To test his method of reduced Bayesian strategies [104], Mikhalevich offered several graduate students to develop, as a practical training, MESM programs that would implement his sequential decision-making algorithm in acceptance sampling problems. Some of these students were O. M. Sharkovskii and N. Z. Shor. This training helped Shor to get a job assignment to the Computer Center of the Academy of Sciences of the USSR after graduating from the Kyiv University in 1958.

In 1958, V. M. Glushkov, a new director of the Computer Center, offered Mikhalevich to head a group of experts in probability theory and mathematical statistics to be engaged in studying the reliability of electronic devices and operations research. That was the beginning of the Kyiv optimization school. At first, Mikhalevich worked in this field together with Bernardo del Rio (a senior lecturer of the Rostov Institute of Railway Transport Engineers, invited by V. M. Glushkov to work in the automation of processes in railway transport) and three young experts Yu. Ermoliev, V. Shkurba, and N. Shor.

From its establishment by Mikhalevich, the Department of Economic Cybernetics was a source of experts in optimal planning, management of the national economy, operations research, design of complex structures and systems, simulation and automation of transport processes, etc. In 1960–1962, more than 100 experts from different regions of the USSR (including Yakutia, Irkutsk, Transcaucasia, Central Asia, and Moldova) were trained at the department; some of them remained to work there.

From the outset, the theoretical developments of the department were motivated by the necessity of solving problems in optimal planning and design.

In developing numerical algorithms to solve some technical and economic extremum problems, Mikhalevich noticed that the ideas of the theory of sequential statistical decisions were useful in analyzing a rather wide class of multivariate optimization problems and in selecting optimal and suboptimal algorithms to solve them. The theory of statistical decisions, including the sequential-search algorithm [101, 103] developed by Mikhalevich, introduces the concept of realization of a trial as a random event statistically related to a set of individual objects with respect to which a decision is sought and its quality is assessed. The realization of a trial results in the redistribution of the probability measures of quantities and attributes that characterize individual objects. The selection of a rational algorithm to search for an acceptable decision is based on sequential design of experiments depending on the previous realizations of trials. In considering deterministic multivariate problems, one may encounter the degenerate yet rather important case where, after realization of trials, some subset of individual objects takes on the measure 0, that is, it can no longer be considered an acceptable variant. This special case was called "algorithm of sequential analysis of variants." It was presented for the first time at the fourth All-Union Mathematical Congress in July 1961 [119] and later in the joint publication of Mikhalevich and Shor [120]. This algorithm quickly became commonly accepted and was widely used by academician N. N. Moiseev and his followers at the Computing Center of the Academy of Sciences of the USSR [126] and by Belarussian Professors V. S. Tanaev and V. A. Emelichev [43, 168].

Mikhalevich's closest followers N. Z. Shor and V.V. Shkurba and their colleagues L. A. Galustova, G. A. Donets, A. N. Sibirko, A. I. Kuksa, V. A. Trubin, and others contributed significantly to the theoretical development of the algorithm of sequential analysis of variants and its application to solve economic and design and engineering problems. Sequential optimization algorithms are also described in the joint monograph by V. S. Mikhalevich and V. L. Volkovich [105]. The paper [139] describes sequential algorithms for solving mixed linear-programming problems with preliminary rejection of redundant constraints and variables that take only zero values in the optimal solution.

Professor V. L. Volkovich and academician V. S. Mikhalevich

A rather comprehensive and concentrated description of the basic ideas behind sequential decision making can be found in [101, 102], which underlie Mikhalevich's Dr.Sci. thesis and in [17] he edited.

From the standpoint of formal logic, the algorithm of sequential analysis of variants can be reduced to repeating the following sequence of operations:

– Divide the sets of possible variants of problem solutions into a family of subsets, each having additional specific properties.
– Use these properties in search for logical contradictions in the description of individual subsets.
– Omit from consideration the subsets whose description is logically contradictory.

The procedure of sequential analysis of variants implies generating variants and selecting operators for their analysis in such a way that parts of variants are rejected once they have been discovered to be unpromising, even if not completely generated. Rejecting unpromising parts of variants also rejects the sets of their continuation and thus saves much computing resources. The more specific properties are used, the greater the savings.

The algorithm of sequential analysis of variants includes the above-mentioned generalization of the well-known Bellman's optimality principle to a wider class of optimization problems.

In what follows, we will formulate this principle and give a methodological procedure free from some constraints inherent in dynamic programming.

Consider some basic set X. Denote by $P(X)$ a set of finite sequences of the form $p = \{x_1, \ldots, x_{K_p}\}$, $x_i \in X$, $1 \le i \le K_p$. Choose some subset of admissible sequences $W(X) \subseteq P(X)$ in this set and a subset of complete admissible sequences $\overline{W} \subseteq W(X)$ in $W(X)$.

Consider a sequence p. Let its l-initial interval be a sequence of the form $p_l(x_1, \ldots, x_l)$, $1 \le l \le K_p$, and its q-final interval be a sequence of the form $p^q = (x_q, x_{q+1}, \ldots, x_{K_p})$, $1 \le q \le K_p$. If $q = l + 1$, then the respective parts of p are called conjugate.

Consider two arbitrary admissible sequences p_1 and p_2. In p_1, choose the l_1-initial interval p_{1l_1} and conjugate final interval $p_1^{l_1+1}$; in p_2, choose the l_2-initial interval p_{2l_2} and conjugate final interval $p_2^{l_2+1}$.

A functional Φ defined on the set $W(X)$ is called monotonically recursive if the truth of $p_{1l_1} \in W$, $p_{2l_2} \in W$, $p_1^{l_1+1} \equiv p_2^{l_2+1}$, and $\Phi(p_{1l_1}) < \Phi(p_{2l_2})$ implies the fulfillment of $\Phi(p_1) < \Phi(p_2)$. Denote $\Phi^* = \sup\{\Phi(p) \,|\, p \in \overline{W}\}$. A sequence $p^* \in \overline{W}$ is called maximum if $\Phi(p^*) = \Phi^*$.

Consider an admissible sequence p. The subset consisting of elements for which p is the initial interval is called a p-generic set. The set of all final intervals of elements of the p-generic set conjugate to p is called the set of continuations $P(p)$. The statement below is true [17].

Theorem 2.1. *Given a monotonically recursive functional Φ and two admissible sequences p_1 and p_2 such that $\Phi(p_1) < \Phi(p_2)$ and $P(p_1) \subseteq P(p_2)$, the elements of the set $P(p_1)$ cannot be maximum sequences.*

Most applications of the algorithm of sequential analysis of variants to be discussed below are based on the generalized optimality principle.

In the 1960–1970s, the Soviet Union actively developed the transport infrastructure, including pipeline networks to export oil and natural gas, created the Unified Energy System, built large industrial facilities (primarily, for the manufacturing and defense industries), and greatly developed air transport. This rapidly expanded the scope of design work and substantially raised the design quality standards, which stimulated the introduction of mathematical design methods. The Institute of Cybernetics, as one of the leading scientific centers, could not keep out of these research efforts. The algorithm of sequential analysis of variants and the generalized optimality principle were applied to solve optimal-design problems for roads [118, 140] and electric and gas networks [121] to find the shortest network routes [6] and critical routes in network diagrams, to solve industrial location problems [118], scheduling problems [117], and some other discrete problems [120].

Let us discuss in more detail some of the classes of optimization models that arose in these applications and the peculiarities of using the mathematics mentioned above.

In the early 1960s, the USSR carried out large-scale projects related to the development of various transportation networks, including power transmission lines and pipelines. Because of the large-scale and limited resources for such projects, it became important to enhance the efficiency of network designs. The complexity of design problems, the necessity of allowing for numerous aspects, and high-quality requirements necessitated the use of mathematical methods to solve such problems.

Problems of designing transport networks implied choosing the network configuration, network components, and their characteristics from a finite set of admissible elements defined by operating engineering standards. The network should be designed to transport a certain product from suppliers to consumers so as to keep the balance between production and consumption and to minimize the total costs of creating and maintaining the network. These requirements can be formalized as an integer optimization problem with recurrent constraints. The objective function and the right-hand sides of the constraints in such a problem are separable functions with linear and nonlinear components. Problems of such type (but less complex) were considered in the fundamental work [11] of R. Bellman, the founder of dynamic programming, which made it expedient to apply the algorithm of sequential analysis of variants to solve them.

Optimization models began to be used to design electric networks in the USSR in the early 1960s. In 1961, a representative of the Ministry of Energy of the USSR organized a relevant meeting and invited representatives of the Institute of Cybernetics (then the Computer Center), Institute of Automation, Polytechnic Institute, and VNDPI "Sil'elektro." Each of the institutes was assigned to develop a technique for the optimization of rural electric networks. The Institute of Cybernetics achieved the most success by focusing on solving the following tasks:

1. Calculation of optimal parameters for 10-kV networks using digital computing machines
2. Computation of optimal parameters for 35-, 10-, and 0.4-kV networks
3. Selection of an optimal configuration of electric networks
4. Selection of optimal parameters considering an optimal arrangement of transformer substations

Computational methods for the first two problems were based on the sequential analysis of variants. For the third problem, L. A. Galustova proposed a heuristic method, which also employed some ideas of the sequential analysis of variants but was not strongly valid mathematically. Nevertheless, this method demonstrated its efficiency during computations based on real data.

These subjects were developed for several years. A series of experimental computations were performed in cooperation with "Sil'elektro," the savings being 10–20 %. The report, together with an implementation and savings certificate, was delivered to the customer and to the Presidium of the Academy of Sciences of the USSR. The above-mentioned methods were also used to design Latvian electric networks (with participation of Ionas Motskus). The results of this work were summarized in the book *A Technique for Optimal Design of Rural Electric Networks* edited by Mikhalevich and in popular science publications.

Almost simultaneously with the above-mentioned research, Gazprom of the USSR together with the Ministry of Energy initiated and guided the automation of gas pipeline design with the participation of the Institute of Cybernetics, VNDI "Gazprom" (Moscow), and VNDPI "Transgaz" (Kyiv).

Several tasks were assigned to the executors.

The first task was to perform a feasibility study for large-diameter gas pipelines, namely, to determine their parameters such as length and diameter and to choose the type and operating modes of engines for gas-compressor stations. The modification of the existing gas pipelines was also an issue: what pipelines should be added to those already laid, what operating conditions of gas pipelines should be, etc. A databank of all the gas pipelines of the Soviet Union was created at the Institute of Cybernetics. Other organizations were also engaged in the design of gas pipelines; however, they tested their algorithms against tentative examples, while the Institute of Cybernetics used real data, which made it possible to refine algorithms by making better allowance for the features of problems.

The second task was to choose optimal parameters of a gas pipeline, taking into account the dynamics of its development. Precedently, the dynamics of investment in the development of gas pipelines and installation of compressors and the possibility of equipment conversion were disregarded. For example, in accomplishing this task, the possibility of using operable overage aircraft engines as compressors was analyzed. A great many engines were expected to arrive after the planned reequipment of the Soviet Army in the 1970s. Not only current but also 5- to 10-year investments were determined. The studies carried out at the Institute of Cybernetics justified that the diameter of gas pipes should be increased from 70 to 100 cm.

These developments were discussed at a top-level meeting in the Kremlin, Moscow. Directors of branch institutes were present. Institute of Cybernetics was represented by L. A. Galustova who demonstrated calculations made at the institute and compared them with American assessment of the development of gas pipelines in the Midwest of the USA. The results appeared similar.

The third major task was as follows: Given the powers of different distributions of gas flows among branches of a gas pipeline, find the optimal distribution for a single gas supply network. The objective function in this problem was nonlinear; to find its extremum, subgradient methods to be discussed in the next section were applied. The calculations were performed for VNDPI "Transgaz" (Kyiv), VNDI "Gazprom" (Moscow), and "Soyuzgazavtomatika." Comparing this project with the previous ones showed that the introduction of optimization modeling saved 50–60 million rubles at 1960 values since the development of the gas-supplying system of the entire Soviet Union was considered. Gazprom refused to sign an act for such an amount since they were afraid that the Academy of Sciences would take away some of the money. As a compromise, an act for 20 million rubles was signed.

This approach was also followed in the 1960–1970s to design oil and product pipelines. As N. I. Rosina (a former employee of the Institute of Cybernetics who took an active part in the above projects) recollects, the algorithm of sequential analysis of variants made it possible to obtain the following results.

1. Mathematical models and computational algorithms were developed to optimize the arrangement of oil pumping stations along oil trunk pipelines and to determine the optimal pipe diameter for separate segments of an oil pipeline, considering the planned (or predicted) volume of oil pumped and the prospects for the development of oil transportation system. These models and algorithms were used to design the Anzhero-Sudzhensk–Krasnoyarsk–Irkutsk trunk pipeline and the oil pipelines that connected Ukrainian oil refineries with the existing oil transportation system.

2. The algorithms were modified to solve problems of arrangement of auxiliary facilities such as cathodic protection stations (with selection of the type of station), equipment of radio relay communication lines, heat tracing stations for high-paraffin oil pipelines, etc. Problems of the arrangement of these objects were solved in combination with the problem of the arrangement of oil pumping stations, considering the possibility of arranging several objects on one site. This resulted in the following synergetic effect: The gain from solving a set of problems exceeded the sum of gains from solving individual problems. The problem of simultaneous arrangement of stations and auxiliary objects had a high dimension, which increased directly proportional to the number of alternative arrangements of each object as new types of objects were added. The total number of alternative arrangements of pipeline elements could reach many billions, that is, a problem of transcomputational complexity arose. Applying the algorithm of sequential analysis of variants reduced the number of alternatives to be analyzed by many orders of magnitude. This made it possible to develop algorithms that took low-performance Soviet computers used in the 1960–1970s a reasonable time to solve such a problem. These algorithms were used, for example, to design pipelines for oil transportation from the Prikaspiiskii region to the European part of the USSR.

3. The algorithm of sequential analysis of variants was used to develop mathematical models and algorithms to design networks of oil pipelines and bulk plants. These algorithms assumed two-stage design. At the first stage, linear optimization problems similar to the transportation problem were solved to assign consumers to bulk plants and to determine their capacity. The solution was then corrected, if possible, to allow for weakly structured aspects disregarded by the model (such as permanent connections between consumers and suppliers). The second stage was to determine the configuration of the oil pipeline network and locations of pumping stations and other infrastructure components as well as oil pipeline diameters. The algorithms used for this purpose were based on the algorithm of sequential analysis of variants. This approach was followed to design oil pipeline networks in Ukraine, Kazakhstan, Byelorussia, and Uzbekistan. The studies were carried out at the "Gipronaftoprovid" design

institute of national standing located in Kyiv in the 1960–1980s. As N. I. Rosina, who worked at this institute, recollects, the software created during the research was successfully used until the early 1990s.

4. Methods of sequential analysis of variants were used at this institute to formulate recommendations on pipeline cleaning frequency and relevant equipment.

Thus, applying the algorithm of sequential analysis of variants made it possible to resolve challenges that arose in designing the transport infrastructure and in identifying the ways of its development.

Along with methods of sequential analysis of variants, decision-support methods were rapidly developed and implemented at the Institute of Cybernetics in the 1960s. They leaned upon the solution of lattice problems, a special class of discrete optimization problems such as the search for critical (shortest and longest) paths, minimum cut, and maximum flow. Solution algorithms for such problems employed Bellman's optimality principle whose relation to the algorithm of sequential analysis of variants was discussed above. It is no accident that network optimization problems drew the close attention of researchers of the Institute of Cybernetics.

In 1963, Mikhalevich was appointed a USSR-wide coordinator of the introduction of network planning and management systems in the field of mechanical engineering, defense, and construction. The results of these studies promoted the implementation of network design systems for managing many large projects and construction of important complex objects.

To search for critical paths, use was made of an algorithm similar to search for the shortest path on a lattice [6] and based on the ideas of sequential analysis of variants.

In 1966, Mikhalevich organized the first All-Union conference on mathematical problems of network planning and management, which initiated many new studies in this field. The conference stimulated large practical developments in the USSR, in which Mikhalevich participated as a scientific adviser. These were applications in the defensive industry (e.g., in planning the launch of the first Soviet nuclear-powered submarine), to support managerial decision making during the construction of petrochemical integrated plants, blast furnaces, sections of Baikal-Amur Mainline (BAM), in stockpile production of shipbuilding yards, etc. Chapter 5 will detail these applications and further development of the methods to solve extremum problems on graphs.

V. V. Shkurba applied the ideas of sequential analysis of variants to solve ordering problems such as optimal arrangement of machine tools in processing centers. These studies gave an impetus to the cooperation with Byelorussian experts in scheduling theory, which resulted in the monograph [168], a classical manual in this field. V. V. Shkurba, who supervised the development of software for solving scheduling problems under limited resources, participated, together with his disciples, in creation of automated production management systems ("Lvov," "Kuntsevo," etc.) [135, 136, 183, 194].

The book [183] analyzes scheduling theory as a division of discrete optimization, classifies relevant problems, and develops exact methods to solve some special problems in scheduling theory and approximate methods to solve the general scheduling problem. Most of these methods were based on the algorithm of sequential analysis of variants. This study was the first to combine methods of applied mathematics and computer technology to solve industrial scheduling and management problems. It also proposed methods for exact solution of some classes of scheduling problems (such as "one machine," "two machine," and "three machine" problems) and methods for approximate solution of more general classes of scheduling problems.

Note that exact methods to solve scheduling problems are efficiently applicable only to some special classes of problems depending on the constraints imposed on the ways of performing elementary operations and on the availability of resources, whose number is no greater than three. It was proved that for $l > 3$, the mathematical complexity of scheduling problems qualitatively increases. The main approach to their solution is to develop approximate methods based on the information simulation of production processes and the wide use of various heuristic methods. Efficient means for the solution of optimization production scheduling problems based on heuristic algorithms, which substantially enhance the quality of production management, were analyzed in the monograph [194]. It was the first to systematize and propose original approaches to estimating the efficiency of heuristic algorithms, which supplement exact methods, including those based on the algorithm of sequential analysis of variants. The ideas of scheduling, modeling of production processes, and management of a large modern industrial enterprise were developed in [136]. This study argued the necessity of implementing systems of management of a modern production association (large industrial enterprise) as multilevel hierarchical systems and outlined the main principles of their development. It systematized the results of improving production planning based on modern economic and mathematical methods, interactive technologies, and principles of system optimization and presented methods and procedure of solving control problems for all the considered levels of hierarchy.

A number of theoretical results related to the quantitative analysis of the complexity of methods of sequential analysis of variants, including network problems of allocation of limited resources, scheduling, and related problems, were published in the monograph [109].

The paper [151] was the first to analyze the class of scheduling problems that arise in the automation of complex production processes such as metal working during galvanization of parts. Approaches to the formalization of such problems and methods of their solution were proposed [152, 154].

These results were used to develop software for the "Galvanik" domestic computer-aided system, which is still widely used in the industry. This system was first launched at the "Arsenal" factory in Kyiv. It was shown many times at international exhibitions and earned gold medals and other high awards. By the way, its authors were awarded with the State Prize of Ukraine in Science and Technology. These results were also used to develop automated production management systems [135, 136, 183, 194]. In the 1970s, the studies were continued by the followers of V. S. Mikhalevich and V. V. Shkurba.

2.2 Solution of Production–Transportation Problems and Development of Methods of Nondifferential Optimization

In 1961, V. S. Mikhalevich came into contact with the Department of Transportation of the Gosplan (State Planning Committee) of the UkrSSR at which A. A. Bakaev (a leading specialist at that time) developed a number of models of integrated transportation of bulk cargoes (e.g., raw material from beet reception centers to sugar mills) whose totality was called *production–transportation problems* and which were of the following form.

Let there be m points of production of some homogeneous product and n points of its consumption. The production of the product unit at each point i $(i = \overline{1, m})$ requires a_{ki} $(k = \overline{1, K})$ *resources of kind* k, where K is the total amount of resources. The manufacturing cost of the product unit at the ith point equals C_i. The total amount of the kth resource B_k of each kind k, the need for the jth product b_j for each point of consumption j $(j = \overline{1, n})$, and the expenditures C_{ij} for the transportation of the product unit from the ith point of production to the jth point of consumption are also known. It is necessary to determine the volume of production x_i at each point i and the volume of product traffic y_{ij} from points of production to points of consumption so that the total cost of production and transportation is as small as possible

$$F = \sum_{i=1}^{m} C_i x_i + \sum_{i=1}^{m} \sum_{j=1}^{n} C_{ij} y_{ij} \rightarrow \min \qquad (2.1)$$

under the resource constraints

$$\sum_{i=1}^{m} a_{ki} x_i \leq B_k, \quad k = \overline{1, K}, \qquad (2.2)$$

and the conditions

$$\sum_{i=1}^{m} y_{ij} \geq b_j, \quad j = \overline{1, n}, \qquad (2.3)$$

of satisfaction of all needs for the product and nonexceeding of the volume of removal of the product over the volume of its production

$$\sum_{j=1}^{n} y_{ij} \leq x_i, \quad i = \overline{1, m}, \qquad (2.4)$$

to which the following conditions of the positiveness of variables of the problem are added:

$$x_i \geq 0, \quad i = \overline{1, \; m}, \tag{2.5}$$

$$y_{ij} \geq 0, \quad i = \overline{1, \; m}, \; j = \overline{1, \; n}. \tag{2.6}$$

Problem (2.1)–(2.6) is a high-dimensional linear-programming problem with weakly filled matrix of constraints. This complicated the use of linear-programming methods, which were well known at that time, in particular, the simplex method, and required the creation of specialized algorithms to solve it. These algorithms were developed proceeding from the following considerations.

Let us consider the optimization problem

$$F_1 = \sum_{i=1}^{m} \sum_{j=1}^{n} C_{ij} y_{ij} \rightarrow \min, \tag{2.7}$$

$$\sum_{i=1}^{m} y_{ij} \geq b_j, \; j = \overline{1, \; n}, \tag{2.8}$$

$$\sum_{j=1}^{n} y_{ij} \leq x_i, \quad i = \overline{1, \; m}, \tag{2.9}$$

$$y_{ij} \geq 0, \quad i = \overline{1, \; m}, \; j = \overline{1, \; n}, \tag{2.10}$$

which is solved for a fixed x_i. It is a transport problem for which efficient algorithms were developed even in the 1940s; in particular, the method of potentials made it possible to simultaneously find solutions to the direct and dual problems. We denote by v_j^* optimal values of dual variables corresponding to constraints (2.8) and by u_i^* values of dual variables corresponding to constraints (2.9). According to the duality theorem, the optimal value of objective function (2.7) is equal to

$$F_1 = \sum_{i=1}^{m} u_i^* x_i - \sum_{i=1}^{n} v_j^* b_j.$$

Thus, instead of problem (2.1)–(2.6), the following problem can be considered:

$$F = \sum_{i=1}^{m} C_i x_i + \sum_{i=1}^{m} u_i^*(x) x_i - \sum_{i=1}^{n} v_j^*(x) b_j$$

under constraints (2.2) and (2.5), where u_j^*, v_j^* is the optimal solution of the problem dual to problem (2.7)–(2.10); the solution is obtained for fixed x_i, $i = \overline{1, \; m}$.

The dimensionality of this problem is essentially smaller; nevertheless, its objective function is nonlinear and non-everywhere differentiable albeit convex. The need for the solution of such problems stimulated works on nonsmooth optimization at the Institute of Cybernetics. In particular, to solve production–transportation problems in network form, N. Z. Shor proposed the method of generalized gradient descent (called subgradient descent later on) in 1961, which is a simple algorithm that makes it possible to minimize convex functions with a discontinuous gradient.

Let us consider an arbitrary convex function $f(x)$ defined on a Euclidean space E^n. Let X^* be a set of minima (it may be empty), and let $x^* \in X^*$ be an arbitrary minimum point; $\inf f(x) = f^*$. A subgradient (a generalized gradient) of the function $f(\bar{x})$ at a point $x \in E^n$ is understood to be a vector $g_f(\bar{x})$ such that the following inequality is fulfilled for all $x \in E^n$:

$$f(x) - f(\bar{x}) \geq (g_f(\bar{x}), \ x - \bar{x}).$$

It follows from the definition of a subgradient that the following condition is satisfied when $f(x) < f(\bar{x})$:

$$(-g_f(\bar{x}), \ x - \bar{x}) > 0.$$

Geometrically, the latter formula means that the antisubgradient at the point $\bar{x}$ makes an acute angle with any straight line drawn from the point $\bar{x}$ in the direction of the point x with a smaller value of $f(x)$. Hence, if the set X^* is nonempty and $x \notin X^*$, then the distance to X^* decreases with the movement from the point $\bar{x}$ in the direction of $- g_f(\bar{x})$ with a rather small step. This simple fact underlies the subgradient method or the method of generalized gradient descent.

The method of generalized gradient descent (GGD) is understood to be the procedure of construction of a sequence $\{x_k\}_{k=0}^{\infty}$, where x_0 is an initial approximation and x_k are computed by the following recursive formula:

$$x_{k+1} = x_k - h_{k+1} \frac{g_f(x_k)}{\|g_f(x_k)\|}, \quad k = 0, \ 1, \ 2, \ldots; \tag{2.11}$$

where $g_f(x_k)$ is an arbitrary subgradient of the function $f(x)$ at a point x_k and h_{k+1} is a step multiplier. If $g_f(x_k) = 0$, then x_k is a minimum point of the function $f(x)$, and the process comes to an end.

The most general result on the convergence of GGD is contained in the following theorem [186].

Theorem 2.2. *Let $f(x)$ be a convex function defined on E^n with a bounded region of minima X^*, and let $\{h_k\}$ ($k = 1, \ 2, \ldots$) be a sequence of numbers that satisfies the conditions*

$$h_k > 0; \quad \lim_{k \to \infty} h_k = 0; \quad \sum_{k=1}^{\infty} h_k = +\infty.$$

Then, for a sequence $\{x_k\}$ $(k = 1, 2, \ldots)$ constructed according to formula (2.11) and for an arbitrary $x_0 \in E^n$, there are the following two possibilities: Some $k = \bar{k}$ can be found out such that we have $x_{\bar{k}} \in X^*$ or

$$\lim_{k \to \infty} \min_{y \in X^*} \|x_k - y\| = 0, \quad \lim_{k \to \infty} f(x_k) = \min_{x \in E^n} f(x) = f^*.$$

The conditions imposed on the sequence $\{h_k\}$ $(k = 1, 2, \ldots)$ in the theorem are well known as classical conditions of step regulation in the GGD. Under definite additional assumptions, GGD variants were obtained that converge at the rate of a geometric progression. For them, a significant part was played by upper bounds of angles between the antigradient direction at a given point and the direction of the straight line drawn from this point to the minimum point.

Further investigations on the development of subgradient methods of nonsmooth optimization were pursued at the Department of Economic Cybernetics of the Institute of Cybernetics. Questions of minimization of non-everywhere differentiable functions [186] arise in solving various problems of mathematical programming when decomposition schemes are used, maximum functions are minimized, exact methods of penalty functions with nonsmooth "penalties" are used, and dual estimates in Boolean and multiextremal problems are constructed. They also arise in solving optimal planning and design problems in which technical and economic characteristics are specified in the form of piecewise-smooth functions.

The method of generalized gradient descent was first used for the minimization of piecewise-smooth convex functions that are used in solving transport and production–transportation problems [189] and then for the class of arbitrary convex functions [116] and convex programming problems. A substantial contribution to the substantiation and development of this method was made by Yu. M. Ermoliev [52] and B. T. Polyak [138], a Moscow mathematician. Ideas of nonsmooth convex analysis were developed in works of B. N. Pshenichnyi, V. F. Demyanov, I. I. Eremin, T. Rokafellar (USA), and other scientists [34, 45, 46, 144, 145, 148]. It should be noted that the interest in the development of subgradient methods in the West was deepened approximately starting with 1974 when investigators saw the key to solving high-dimensional problems in them [213].

However, it turned out that there is a wide range of functions for which the convergence of the subgradient method is weak. This is especially true for the so-called ravine functions. The reason is that the antigradient direction makes an angle with $-(x - x^*)$ that is close to $\frac{\pi}{2}$ and the distance to X^* changes insignificantly with movement in this direction. "Ravine" functions occur rather often in applied problems.

Let us consider, for example, the "ravine" function $F(x) = x_1^2 + 100x_2^2$. After the replacement of the variables $y_1 = x_1$ and $y_2 = 10x_2$, we obtain the new function $F(y) = y_1^2 + y_2^2$, which is very handy for applying the gradient method (the above-mentioned angle is always equal to zero for it). Note that, in most cases, such an obvious replacement is absent. Moreover, some replacement that can improve properties of the function in the vicinity of a point x can worsen them outside of this vicinity. But, in solving a problem with the help of the gradient method, there may be some points at which some replacement will worsen such a function. Thus, it is necessary to develop a method that would allow one to construct a new coordinate system for each iteration and, in case of need, to quickly come back to the initial system. To this end, linear nonorthogonal transformations can be used such as the operation of dilation of an n-dimensional vector space.

Let $x \in E^n$ be an arbitrary vector, and let ξ be an arbitrary direction, $||\xi|| = 1$. Then, the vector x can always be represented in the form

$$x = \gamma(x, \ \xi)\xi + d(x, \ \xi),$$

where $\gamma(x, \ \xi)$ is some scalar quantity and $(d(x, \ \xi), \ \xi) = 0$.

The operator of dilation of a vector x in the direction ξ with a coefficient $\alpha > 0$ is understood to be as follows:

$$R_\alpha(\xi)x = \alpha\gamma(x, \ \xi)\xi + d(x, \ \xi) = \alpha(x, \ \xi)\xi + x - (x, \ \xi)\xi = x + (\alpha - 1)(x, \ \xi)\xi.$$

This operation changes the part $\gamma(x, \ \xi)\xi$ (collinear with ξ) of the vector x by a factor of α and does not change the part $d(x, \ \xi)$ (orthogonal to ξ) of this vector. If this operation is applied to all the vectors that form the space E^n, then we say that the space is extended with a coefficient α in the direction ξ.

For "ravine" functions, the direction of the generalized gradient is almost perpendicular to the axis of the "ravine" at the majority of points. Therefore, the space dilation in the direction of the generalized gradient decreases the "raviness" of the objective function. This implies the following general idea of the method: Its own new coordinate system is used at each iteration. In this system, the generalized gradient of the objective function is calculated, one iteration of the method of generalized gradient descent is performed, and the coordinate system is changed by space dilation in the direction of the generalized gradient. This idea was embodied in numerous space dilation algorithms, some of which will be considered below.

If an algorithm is designed for the solution of a wide circle of problems, then the researcher can choose a mathematical model that most adequately and economically reflects an actual process. With a view to extending the circle of efficiently solved nonsmooth optimization problems, N. Z. Shor proposed in the 1969–1971s and, together with his disciples, experimentally investigated [184, 190] methods of subgradient type with space dilation that are used to solve essentially ravine problems. An algorithm that was developed at that time and dilated the space in the direction of the difference of two successive subgradients (the so-called r-algorithm) remains one of the most practically efficient procedures of nonsmooth optimization even at the present time.

Let us consider this method in more detail. We illustrate it by the example of the problem of minimization of a convex function $f(x)$ defined on E^n. We assume that $f(x)$ has a bounded region of minima X^* so that we have $\lim_{x \to \infty} f(x) = +\infty$.

We choose an initial approximation $x_0 \in E^n$ and a nonsingular matrix B_0 (a unit matrix I_n or a diagonal matrix D_n that has nonnegative elements on its diagonal and with the help of which variables are scaled are mostly chosen).

The first step of the algorithm is made according to the formula $x_1 = x_0 - h_0\eta_0$, where $\eta_0 = B_0B_0^T g_f(x_0)$, h_0 is some step multiplier chosen under the condition that there exists a subgradient $g_f(x_1)$ at the point x_1, and it is such that we have $(g_f(x_1), \eta_0) \leq 0$. When $B_0 = I_n$, we have $\eta_0 = g_f(x_0)$, and the first step coincides with the iteration of the subgradient method with a constant metric.

Let the values of $x_k \in E^n$ and matrices B_k of dimension $n \times n$ be obtained as a result of k $(k = 1, 2, \ldots)$ steps of the process of computations. We describe the $(k + 1)$th step of the extremum searching process:

1. Compute the following quantities: $g_f(x_k)$, that is, the subgradient of the function $f(x)$ at the point x_k, and $r_k = B_k^T(g_f(x_k) - g_f(x_{k-1}))$, that is, the vector of the difference between two successive subgradients in the transformed space.

   The passage from the initial space to the transformed one is specified by the formula $y = A_k x$, where $A_k = B_k^{-1}$. We define the function $\varphi_k(y) = f(B_k y)$ and obtain $g_{\varphi_k}(y) = B_k^T g_f(x)$. Thus, r_k is the difference of two subgradients of the function $\varphi_k(y)$ that are computed at the points $y_k = A_k x_k$ and $\tilde{y}_k = A_k x_{k-1}$.
2. Compute $\xi_k = r_k/\|r_k\|$.
3. Specify the quantity β_k inverse to the space dilation coefficient α_k.
4. Compute $B_{k+1} = B_k R_{\beta_k}(\xi_k)$, where $R_{\beta_k}(\xi_k)$ is the space dilation operator at the $(k + 1)$th step. We note that $B_{k+1} = A_{k+1}^{-1}$.
5. Find $\tilde{g}_k = B_{k+1}^T g_f(x_k)$, that is, the subgradient of the function $\varphi_{k+1} = f(B_{k+1}y)$ at the point $y_{k+1} = A_{k+1}x_k$.
6. Determine $x_{k+1} = x_k - h_k B_{k+1}\tilde{g}_k/\|\tilde{g}_k\|$. This step of the algorithm corresponds to the step of the generalized gradient descent in the space transformed under the action of the operator A_{k+1}.
7. Pass to the next step or terminate the algorithm if some termination conditions are fulfilled.

The practical efficiency of the algorithm depends in many respects on the choice of the step multiplier h_k. In the r-algorithm, h_k is chosen under the condition of approximate search for the minimum of $f(x)$ along the descent direction, and in this case, during the minimization of convex functions, the condition $h_k \geq h_k^*$ should be satisfied, where h_k^* is the value of the multiplier corresponding to the minimum along the direction. Hence, it is necessary that the angle made by the subgradient direction at the point x_{k+1} with the descent direction at the point x_k be not blunt.

In minimizing nonsmooth convex functions, the following variants of the algorithm turned out to be most successful. Space dilation coefficients α_k are chosen equal to 2–3, and an adaptive method of regulation is used for the multiplier step h_k. Some natural number m and constants $q > 1$ and $t_0^0 > 0$ are specified (after k steps, the latter quantity will be accordingly denoted by t_k^0). The algorithm moves from the point x_k in the descent direction with the step t_k^0 until the termination condition for the descent along the direction is fulfilled or the number of steps becomes equal to m. The descent termination condition can be as follows: The value of the function at the next point is no smaller than its value at the previous point; another variant of this condition is as follows: The derivative along the descent direction at a given point is negative. If m steps have been executed and the descent termination condition is not fulfilled, then $t_k^1 = qt_k^0$ is stored instead of t_k^0, where $q > 1$ and the descent continues in the same direction with a larger step. If, after the next m steps, the descent termination condition is not fulfilled, then we take $t_k^2 = qt_k^1$ instead of t_k^1, etc. By virtue of the assumption $\lim_{\|x\| \to \infty} f(x) = +\infty$, after a finite number of steps in the fixed direction, the descent termination condition will be necessarily fulfilled. The step constant $t_k^{p_k} = q^{p_k} t_k^0$ ($p \in \{0, 1, 2, \ldots\}$) that has been used at the last step is considered to be initial for the descent from the point x_{k+1} along a new direction, that is, we have $t_k^0 = t_k^{p_k}$.

As is shown by numerous computational experiments and practical calculations, in most cases when $\alpha \in [2, 3]$, $m = 3$, and h is controlled with the help of the above-mentioned method, the number of steps along a direction, on the average, seldom exceeds 2, and in this case, during n steps of the r-algorithm, the accuracy with respect to the corresponding function is improved, as a rule, by a factor of 3–5.

In 1976–1977, A. S. Nemirovskii and D. B. Yudin [128] and N. Z. Shor [185] independently proposed the ellipsoid method that combines ideas of the cutting-plane and space transformation methods. In fact, this method is a special case of an algorithm of subgradient type with space dilation, which was proposed by N. Z. Shor as long ago as 1970 and in which the following parameters are used: The space dilation coefficient is chosen invariable and equal to

$$\alpha_{k+1} = \alpha = \sqrt{\frac{n+1}{n-1}},$$

and the step control is performed according to the following rule:

$$h_1 = \frac{r}{n+1}; \quad h_{k+1} = h_k \frac{n}{\sqrt{n^2 - 1}}; \quad k = 1, 2, \ldots,$$

where n is the space dimension and r is the radius of a sphere whose center is at the point x_0 and that contains the point x^*. The convergence rate of the ellipsoid method coincides with that of a geometric progression as to the deviation of the best reached value of $f(x)$ from optimal and, at the same time, the denominator of the geometric progression asymptotically depends only on the space dimension n,

$$q_n \approx 1 - \frac{1}{2n^2}.$$

The ellipsoid method received wide popularity all over the world after L. G. Khachiyan [175] proposed (in 1979) a polynomial algorithm based on this method for the solution of linear-programming problems with rational coefficients, which conditioned its wide applications in the complexity theory for optimization algorithms. A number of modifications of the ellipsoid method were proposed in [19]. V. S. Mikhalevich supported the development of the line of investigation into nonsmooth optimization in every possible way and propagandized the results obtained at the department. These works received world recognition mainly owing to him. V. S. Mikhalevich and N. Z. Shor participated in the work of the 11th International Symposium on Mathematical Programming in 1982 at which they presented their plenary reports concerning the development of methods and a technology of solution of high-dimensional optimization problems.

V. M. Glushkov, L. V. Kantorovich, and V. S. Mikhalevich made many efforts for embedding the system of optimum workload of pipe-rolling enterprises of the USSR and shop order management that was created together with the All-Union Research and Development Institute (VNII) for the Pipe Industry (the laboratory headed by A. I. Vainzof). The mathematical core of this system consisted of production–transportation models and nonsmooth optimization methods [116] developed at the Department of Economic Cybernetics at the Cybernetics Institute of the Academy of Sciences of the UkrSSR. V. S. Mikhalevich made a substantial contribution to the embedding of methods of solution of production–transportation problems for optimal planning in civil aviation [15, 122]. Algorithms for the solution of production–transportation problems were realized in the form of application packages (APs), in particular, in the form of the AP PLANER [113].

Numerous applications of nonsmooth optimization in production–transportation problems are considered in [116, 186, 193, 220]. Based on them, the software support was developed for some models of optimal planning, design, and control. For example, high-dimensional production–transportation problems were solved with a view to distributing ships among inland waterways. Problems of a particularly high dimensionality arose in the workload planning for rolling mills. In these problems, the number of orders (to which corresponded consumers of products in models) reached 10,000, the number of states (manufacturers of products) exceeded 50, and the nomenclature consisted of about 1,000 names of kinds of products. Thus, the total number of variables of a multi-index analogue of model (2.1)–(2.6) was about hundreds of millions. Nevertheless, the approach developed above to decomposition made it possible to solve such problems of transcomputational complexity on second-generation computers. (On a computer of the type M-220 available at that time, the process of solution required about 12–15 h to ensure the accuracy about 10^{-3}). Algorithms proposed on this basis were used in planning the workload of rolling mills in Soyuzglavmetal of the USSR Gossnab (State Supplies) with considerable economic effect. Programs that realized methods of nondifferential optimization were also used in creating a system of automated planning of production and shop order management in ferrous metallurgy and for solving problems of distribution of coked coal among by-product coke plants. Together with the State Research and Development Institute of Civil Aviation of the Ministry of Civil Aviation of the USSR, a system of economic–mathematical models and optimization algorithms was developed that used subgradient methods for systems of short-term and long-term planning in civil aviation. They were supported by software tools, and calculations of the prospects of development of the fleet of civil aviation were performed with their help.

A meaningful experience had been accumulated before 1970 in solving other structured problems (similar to the production–transportation problem) of linear and nonlinear programming with the use of subgradient methods, for example, multicommodity network flow problems with bounded capacities, problems of choice of an optimal structure of machine and tractor fleets in individual farms and in a branch on the whole, etc. This experience had shown that methods of nondifferential optimization are most efficient in combination with various decomposition schemes in solving coordination problems of the form (2.7)–(2.10).

It should be mentioned that groundworks on production–transportation problems can be efficiently used now in logistics systems of large mining and smelting and machine-building holdings.

Works on the application of nonsmooth optimization methods to the solution of economic problems also continued in the Cybernetics Institute of AS of UkrSSR throughout the 1970s and 1980s. In particular, under the scientific leadership of V. S. Mikhalevich, new numerical methods were developed that were used for the solution of special classes of nonlinear-programming problems [191, 193, 221].

The majority of such problems are multiextremal and can be exactly solved only with the use of exhaustive search methods (e.g., the branch-and-bound method). At the same time, to increase the practical efficiency of such methods, it is expedient to construct more exact lower estimates of objective functions in minimization problems. N. Z. Shor suggested to use dual quadratic estimates for some multiextremal and combinatorial optimization problems that, as a rule, are more exact in comparison with corresponding linear estimates and can be obtained using nonsmooth optimization methods. Moreover, he developed the methodology of generation of functionally redundant quadratic constraints whose introduction into a model does not change the essence of a problem but nevertheless allows one to obtain more accurate dual estimates in some cases. Examples of functionally redundant constraints can be as follows:

(a) Quadratic consequences of linear constraints, for example, a quadratic constraint in the form $(b_i^T x + c_i)(b_j^T x + c_j) \geq 0$ is a consequence of the following two linear inequality constraints: $b_i^T x + c_i \geq 0$ and $b_j^T x + c_j \geq 0$.

(b) Quadratic constraints that characterize the ambiguity of representation of the product of three, four, or larger numbers of variables of a problem. As a rule, they take place in reducing a polynomial problem to a quadratic one. For example, we have variables $x_1, x_2 = x_1^2$, and $x_3 = x_1^3$. Then, the quadratic constraint in the form of the equality $x_2^2 - x_1 x_3 = 0$ is a consequence of an ambiguous representation of x_1^4, namely, $x_1^4 = (x_1^2)^2 = (x_1^3)(x_1)$.

(c) Quadratic constraints that are consequences of the Boolean or binary nature of variables of a problem. For example, for binary variables $x_i^2 = 1$, $x_j^2 = 1$, and $x_k^2 = 1$, the following quadratic inequality is always satisfied $x_i x_j + x_i x_k + x_j x_k \geq -1$.

The idea of functionally redundant constraints has played the key role in analyzing problems of searching for the global minimum of a polynomial $P(z_1, \ldots, z_n)$ in several variables. It turned out that, in using a specially constructed equivalent quadratic problem, the corresponding dual estimates coincide with the minimal value P^* of such a polynomial if and only if the nonnegative polynomial $\bar{P}(z) = P(z) - P^*$ is presented in the form of the sum of squares of other polynomials [187, 188]. This result turned out to be directly connected with the 17th classical Hilbert problem on the representation of a nonnegative rational function in the form of the sum of squares of rational functions. The mentioned investigations extend the domain of efficiently solved global polynomial minimization problems including convex minimization problems as a particular case. Algorithms for construction of dual quadratic estimates were proposed for extremum problems on graphs such as problems of searching for maximal weighted independent set of graphs, minimum graph-coloring problems [193, 219, 220], the combinatorial optimal graph partition problem [193, 220], the weighted maximum cut problem on graphs, etc. Based of nonsmooth optimization methods, new algorithms were developed for the construction of volume-optimal inscribed and circumscribed ellipsoids [221], algorithms for the solution of some problems of the stability theory for control over dynamic systems, etc. These investigations are rather completely presented in [220].

Nondifferential optimization methods played an important role in creating packages of applied programs. Theoretical investigations in the field of construction of automated software tools began in the 1960s and had led to the development of complicated interactive software systems destined for the solution of different classes of optimization problems. In this line of investigations, important results were obtained in different years by E. L. Yushchenko, P. I. Andon, A. A. Letichevskii, V. N. Redko, I. N. Parasyuk, O. L. Perevozchikova, and other researchers. These results are presented, in particular, in [23, 113–115, 134]. Programs that use subgradient methods are included in the APs PLANER, DISPRO, and DISNEL created at the Institute of Cybernetics in the 1980s for ES EVM computers. These packages realized a wide spectrum of methods for the solution and investigation of problems of optimal planning, design and management, distribution and renewal of enterprises, design of technical devices and machines, and work scheduling under bounded resources. In [114], a review of optimization methods is presented that were developed in the Institute of Cybernetics and became central in realizing the mentioned packages of applied programs. In [113], mathematical models of applied problems and the system software of the AP PLANER are described. The majority of these models were included in an improved form in the AP DISNEL. In [115], the destination, classes of solved problems, and system and algorithmic supports for the AP DISPRO-3 are described. Here, nonsmooth optimization methods were mainly used in solving "evaluative" problems using the branch-and-bound method for special classes of discrete problems.

Important results connected with the investigation and solution of some special problems of discrete optimization were obtained by Professor V. A. Trubin who worked many years under the leadership of V. S. Mikhalevich at the Department of Economic Cybernetics. He studied properties of the polytope M of the bounded combinatorial partition problem

$$M = \{Ax = 1, \ x \leq 0\},$$

where A is a (0,1) matrix and 1 and 0 are the unit and zero vectors of the corresponding dimension.

Let X be the set of vertices of the polytope M, and let R be the set of its edges. Together with the polytope $M = M(X, R)$, we consider a polytope $\overline{W} = M(\overline{X}, \overline{R})$ generated by only integer-valued vertices $\overline{X} \subset X$ with the set of edges $\overline{R}$. In 1969, V. A. Trubin [171] proved the following property of $\overline{M}$: $\overline{R}$ is a subset of R. This property allows one to basically modify the simplex method of linear programming to solve problems of integer-valued optimization on the polytope M. Its modification lies in the prohibition of any movement from the current feasible solution x to the adjacent (along an edge from R) non-integer-valued solution y. In other words, transitions only to adjacent integer-valued vertices are allowable if the value of the criterion of the problem is improved in this case. The mentioned property directly implies the proof of the well-known Hirsch hypothesis on the considered class of integer-valued optimization problems. Thus, this property allows one to construct an algorithm for the solution of the partition problem with a rather small number of iterations. However, as was almost simultaneously noted by many researchers, the amount of computations required for the execution of one iteration of this algorithm can turn out to be rather computationally intensive, which is connected with a high degeneracy of the polytope $\overline{M}$.

Many optimization problems on graphs and networks belong to the class of partition problems such as the search for a maximum independent set, minimal covering, coloring, problem of distribution and synthesis of communication networks, and problems of cargo transportation and scheduling.

Investigations in the field of minimization of concave functions over a transportation polytope [116, 173] began as long ago as 1968. This class of problems includes numerous extremal problems that arise in the design of communication and information networks, distribution of production, choice of the configuration of the equipment for multiproduct manufactures, and unification and standardization. One of the most promising approaches to the solution of these problems is their reduction to problems of discrete optimization of a special structure and the development of decomposition methods that maximally take into account the structural properties of the latter problems during solving them. A general decomposition scheme was developed for the construction of double-ended estimates of functionals in the mentioned problems. It includes a general oracle algorithm of linear optimization (based on the scheme of the simplex method or on nonsmooth optimization methods) in which the role of an oracle that determines the direction of movement at a given point (the subgradient or its projection) and step size is played by minimum-cut and shortest path procedures applied to some graph constructed for a given step. The construction of the decomposition approach allows for the introduction of a partial order over the columns of the constraint matrix of the problem being considered, which, in turn, allows one to replace an algorithm of exact construction of estimates by its approximate variant of gradient type. This replacement dramatically reduces the computational intensity of construction of estimates but, unfortunately, can worsen them. However, numerous experiments with various practical problems showed that the gap between these estimates remains small, and the time required for the solution of problems is completely acceptable [116]. An approximate approach is often unique in connection with high dimensions of problems that arise in design practice. It was used for the solution of problems of distribution of locomotive and car repair bases, design of water supply systems in the Zaporizhia and Zakarpattia oblasts of Ukraine, municipal communication and electric networks, ship's pipeline networks, problems of choosing configurations of tube-rolling shops, etc.

Over the past 10 years, methods of nondifferential optimization have found a number of important practical applications. To solve special classes of problems of two-stage stochastic programming with simple and fixed recursions, the programs Shor1 (whose authors are N. Z. Shor and A. P. Lykhovyd) and Shor2 (whose authors are N. Z. Shor and N. G. Zhurbenko) were developed. The programs are based on the use of the scheme of decomposition with respect to variables and solution of a nonsmooth coordination problem with the help of the r-algorithm (which is one of the most well-known subgradient methods) with adaptive step regulation. The Shor1 and Shor2 programs are introduced into the system SLP-IOR developed at the Institute for Operations Research (Zurich, Switzerland) for modeling stochastic linear-programming problems. A software support was also developed for some problems of optimum design and network routing with allowance for possible failures of some components of a network and a modification in requirements on flows. Its description is included in [192]. Methods of nonsmooth optimization were used to solve some problems of design of power installations [89]. These works were performed by Yu. P. Laptin and M. G. Zhurbenko, and their results were introduced into the Kharkiv Central Design Bureau "Energoprogres."

The mentioned methods were also used by Professor E. M. Kiseleva who (together with her disciples) developed the mathematical theory of continuous problems of optimal partition of sets in an n-dimensional Euclidean space that, first, are nonclassical problems of infinitely dimensional mathematical programming with Boolean variables and, second, form one more source of generation of nonsmooth optimization problems. Based on the proposed and theoretically substantiated methods for solving problems of optimal partition of sets [77], algorithms are constructed whose component is the r-algorithm of N. Z. Shor and its modifications.

In [116, 169, 170, 172, 173], the results of investigations are presented that are devoted to the development of polynomial algorithms for the solution of some network analysis and synthesis problems. Here, we are dealing with the problem of distribution of production over treelike networks, Weber problems in a space with the L_1-metric, problems of determination of the strength of a network and its optimal strengthening, and also problems of packing and covering of a network by spanning trees and branchings in continuous and integer-valued statements.

The further advancement of works on the mentioned subjects took place in the 1990s when the Institute of Cybernetics performed joint investigations with several leading companies, in particular, with the well-known Japanese Wacom Co., Ltd., on design methods for telecommunication networks. A distinctive feature of solving these problems was the simultaneous application of the methods of sequential analysis of variants (SAV) and nondifferential optimization algorithms mentioned above.

In 1993–1994, the method of sequential analysis of variants was used in solving problems of planning optimal fueling regimes of thermal power stations over the daily and week hourly scheduled period. This project was headed by Masaki Ito from Wacom Co., Ltd. and its research management was carried out by N. Z. Shor and V. A. Trubin from the Institute of Cybernetics. The mentioned problem was reduced to the following integer-valued problem: It is required to find

$$\min \sum_{i=1}^{n} q_i(x_i), \tag{2.12}$$

under the following constraints:

$$\sum_{i=1}^{n} x_i = b, \tag{2.13}$$

$$x_i \in \{0 \vee [l_i, \ u_i]\}, \quad i = 1, \ 2, \ldots, \ n. \tag{2.14}$$

Here, $b, l_i, u_i, \ i = 1, \ 2, \ldots, \ n$, are integers such that $b > 0$, $l_i \geq 0$, and $u_i > l_i$ and $q_i(x_i)$ are nonnegative bounded functions defined over an interval $[l_i, \ u_i]$, $i = 1, \ldots, \ n$.

In the general case, for large b, problem (2.12)–(2.14) is rather complicated, and it is inexpedient to use classical methods (e.g., necessary extremum conditions) to solve it. Nevertheless, the fact that the objective function is separable allows one to apply the method of sequential analysis of variants that generalizes the idea of dynamic programming [11, 17].

For problem (2.12)–(2.14), the system of functional Bellman equations is of the form

$$f_1(y) = \begin{cases} 0, & y = 0, \\ q_1(y), & l_1 \leq y \leq u_1, \\ \infty, & 0 < y < l_1, \ u_1 < y \leq b, \end{cases} \tag{2.15}$$

$$f_k(y) = \min_{l_k \leq x \leq u_k} \{f_{k-1}(y), \ q_k(x) + f_{k-1}(y - x)\}, \quad k = 2, \ldots, \ n, \tag{2.16}$$

where $0 \leq y \leq b$ and $f_k(y)$ are Bellman functions, $k = 1, \ldots, \ n$.

In what follows, we call the algorithm constructed according to relationships (2.15) and (2.16) Algorithm 1. Its distinctive feature is that it does not depend on the form of functions $q_i(x_i)$, $i = 1, \ldots, \ n$. The basic operation for Algorithm 1 is the addition operation. It determines its complexity (the number of additions in Algorithm 1 is bounded by the value of $b \sum_{i=1}^{n} (u_i - l_i)$). In the case of piecewise concave functions, the speed of Algorithm 1 can be increased with simultaneously decreasing the number of additions. To this end, the lemma presented below is used.

Lemma 2.1. *Let* $x^* = (x_1^*, \dots, x_n^*)$ *be an optimum solution of problem (2.12)–(2.14) for pieces–concave functions* $q_i(x_i)$, $i = 1, \dots, n$. *Then, one of the following statements is true*:

(a) *If an optimum solution* x^* *is unique, then it contains no more than one variable that assumes its value in the interval of concavity. All the other variables assume their values at the end points of intervals of concavity.*

(b) *If an optimum solution is not unique, then there is an optimum solution that contains no more than one variable whose values are within the interval of concavity.*

Let us pay attention to analogies between the problem being considered and linear-programming problems. A distinctive feature of the structure of optimum solutions (the fact that there is at least one basic solution with a relatively small amount of nonzero components is among such solutions) in the latter problems made it possible to create efficient exhaustive algorithms among which the simplex method is most well known. The lemma presented above also allows one to apply a similar approach to the solution of the problem of planning optimal regimes of fueling power-generating units.

Using properties of an optimum solution, one can "accelerate" Algorithm 1 if functions $q_i(x_i)$, $i = 1, \dots, n$, are piecewise concave. For the variables x for which a conditionally optimum solution of the $(k - 1)$th function $f_{k-1}(y)$ already contains a variable whose values belong to the interval of concavity, it is sufficient to take into account only the ends of intervals of concavity of the kth function $q_k(x_k)$, $l_k \leq x_k \leq u_k$, in constructing the function $f_k(y)$.

We call the algorithm modified in this manner Algorithm 2. Its computational scheme practically repeats that of Algorithm 1. A distinction is that the modified algorithm retains an array of flags that indicate whether the current value of $f_{k-1}(y)$ is reached at the ends of intervals of concavity for all $1, \dots, k - 1$ variables or some of these variables have already assumed a value in the interval of concavity. Depending on the value of such a flag, the loop for recomputation of $f_k(y)$ is executed for all points of the interval $[l_k, u_k]$ or only for the end points of intervals of concavity of functions q_k.

We estimate the computational complexity of Algorithm 2. Here, the number of addition operations is about nbN, where N is the maximal number of intervals of concavity, which provides a significant computational effect, especially when $N << \max_i (u_i - l_i)$. Moreover, the additional array of flags makes it possible to find out whether the minimum of the objective function is reached or not reached at the ends of intervals of concavity. This provides an optimum solution to problem (2.12)–(2.14) with an additional useful characteristic in the case of piecewise concave functions.

It should be noted that computational experiments performed on some test problems have shown doubtless advantages of Algorithm 2. There is reason to hope that the time required for the solution of real-world problems of large sizes will be acceptable if this algorithm is used together with decomposition schemes.

The operating speed of the SAV method plays here a decisive role since it is used in computing subgradients of nonsmooth functions at every step of the subgradient process.

The methods of nondifferential optimization that are developed at the V. M. Glushkov Institute of Cybernetics were also used to solve problems of optimal design of structures of reliable networks within the framework of investigations pursued according to a project of the Science and Technology Center in Ukraine (STCU). In particular, the developers of the project investigated problems of designing a minimum-cost network under failure conditions of its separate units, finding throughputs of edges in a reliable directed network, designing an optimal logical structure of a reliable network, modernizing a reliable network, and optimizing networks with allowance for incomplete information and also problems of forward planning of shipping operations and finding of an optimal nomenclature for a rolling stock. To search for optimum solutions under these conditions, they used methods of nondifferential optimization, methods of local discrete search with elements of sequential analysis of variants, and decomposition schemes [192].

As an example, let us consider two mathematical models presented in [192], namely, the problem of finding edges throughputs in a reliable directed network with transmission of flows along arbitrary paths and with transmission of flows along a given set of admissible paths. In both cases, it is necessary to provide the adequate functioning of the network under failure conditions (its separate components, namely, edges and nodes, can fail).

A network failure is understood to be a network state that decreases the throughput of one or several of its edges. In formulating a mathematical model of the problem of finding the throughput of edges of a reliable directed network with transmission of flows along arbitrary paths, we use the following data:

1. A directed network $N(V,A)$ is specified by a set of nodes V and a set of edges A. For an edge $(i,j) \in A$, we denote by c_{ij} the cost of creation of unit throughput and by y_{ij}^0 the available throughput resource.

2. A set of requests D for volumes of flows between pairs of nodes from some subset $V_0 \subset V$ is given. Each element of the set D is specified by the following three numerical quantities: a pair (r, s) and the corresponding volume of the flow d_{rs} that should be allowed to pass through the network from its node $r \in V_0$ (source) to its node $s \in V_0$ (drain). For all pairs (r, s), the corresponding values of volumes d_{rs} are given in the same units as throughput capacities of edges.

3. A set T of possible failures of the network $N(V,A)$ is given. A failure $t \in T$ of the network is characterized by a collection of coefficients $0 \le \mu_{ijt} \le 1$ for all $(i,j) \in A$, where a concrete coefficient μ_{ijt} testifies to the fact that the throughput of an edge (i,j) in the network decreases by a factor of $1/\mu_{ijt}$ as a result of the failure t. For the convenience of description of mathematical models, we will conditionally consider that the index $t = 0$ as the "zero" failure of the network $N(V,A)$ to which corresponds the collection of coefficients $\mu_{ij0} = 1$ for all $(i,j) \in A$. In fact, the "zero" failure of the network means that the network $N(V,A)$ functions as in the case when failures are absent.

Let us make the following assumption on the input data for items (1)–(3).

Assumption 2.1. *For the directed network $N(V, A)$, all the requests D for the transmission of volumes of flows between pairs of nodes can be satisfied for unbounded values of throughputs of edges of this network $N(V, A)$.*

Let $Y = \{y_{ij}, (i, j) \in A\}$ be the set of unknown values of throughputs of edges $(i, j) \in A$ that should be added to the already existing values of throughputs of edges $Y^0 = \{y_{ij}^0, (i, j) \in A\}$ of the network, and let x_{ijt}^{rs} be an unknown value of the portion of the volume of the flow d_{rs}, $(r, s) \in D$ that will pass through the edge $(i, j) \in A$ after the tth network failure. Then the mathematical model of the problem of finding values of Y with "optimal total cost" for the reliable directed network $N(V, A)$ can be formulated as the following problem of mathematical programming: It is necessary to minimize the linear function

$$\sum_{(i, j) \in A} c_{ij} y_{ij} \tag{2.17}$$

under the constraints

$$\sum_{(r, s) \in D} x_{ijt}^{rs} \leq \mu_{ijt}(y_{ij}^0 + y_{ij}), \quad t \in (0 \cup T), \quad (i, j) \in A, \tag{2.18}$$

$$\sum_{j:(i,j) \in A} x_{ijt}^{rs} - \sum_{j:(j,i) \in A} x_{jit}^{rs} = \begin{cases} d_{rs}, & i = r, \\ 0, & i \neq r, s, \\ -d_{rs}, & i = s, \end{cases} \quad t \in (0 \cup T), \quad (r, s) \in D, \tag{2.19}$$

$$x_{ijt}^{rs} \geq 0, \quad t \in (0 \cup T), \quad (i, j) \in A, \quad (r, s) \in D, \tag{2.20}$$

$$y_{ij} \geq 0, \quad (i, j) \in A. \tag{2.21}$$

Here, function (2.17) being minimized determines the total expenditures with allowance for the cost of the increase in the throughput of edges that should be added to the existing edges with a view to ensuring the reliability of the network $N(V, A)$.

If the transmission of flows should be restricted to a given set of admissible paths, then the following item is added to conditions (1)–(3):

4. A set of admissible paths in a network $P = \bigcup P(r, s)$ is given, where $P(r, s)$ is the subset of paths in the network $N(V, A)$ that connect a source r with drains s, $(r, s) \in D$ and through which (and only through which) a flow can be transmitted from the node r to the node s. We assume that such collections of paths are given for all pairs $(r, s) \in D$. We will specify a concrete path $P_k(r, s) \subseteq P(r, s)$, that is, the path with a number k for transmission of a volume of the flow from the node r to the node s, by a vector a_k^{rs} whose length equals $|A|$, which consists

of zeros and unities, and in which unities correspond to the edges of the network $N(V, A)$ through which this path passes and zeros correspond to the edges that do not belong to this path.

Assumption 2.2. *The set P contains nonempty subsets $P(r, s)$ for all pairs $(r, s) \in D$.*

For the directed network $N(V, A)$, a new problem of finding values of throughputs of edges $Y = \{y_{ij}, (i, j) \in A\}$ with minimum total cost that should be added to the already existing edges (Y_0) in order that the network $N(V, A)$ become reliable can be formulated as follows: minimize the linear function

$$\sum_{(i,j) \in A} c_{ij} y_{ij} \tag{2.22}$$

under the constraints

$$\sum_{(r,s) \in D} \sum_{k \in P(r,s)} a_k^{rs} x_{kt}^{rs} \leq \mu_{ijt}(y_{ij}^0 + y_{ij}), \quad t \in (0 \cup T), \quad (i, j) \in A, \tag{2.23}$$

$$\sum_{k \in P(r,s)} x_{kt}^{rs} = d_{rs}, \quad t \in (0 \cup T), \quad (r, s) \in D, \tag{2.24}$$

$$x_{kt}^{rs} \geq 0, \quad t \in (0 \cup T), \quad (r, s) \in D, \quad k \in P(r, s), \tag{2.25}$$

$$y_{ij} \geq 0, \quad (i, j) \in A, \tag{2.26}$$

where x_{kt}^{rs} is an unknown value of the subvolume of the flow d_{rs} that will pass through the kth path, $k \in P(r, s)$, in the case of the tth failure of the network.

A model destined for the solution of problems of forward transportation planning and rational allocation of appropriations for the reconstruction of transport (railway, automobile, and aviation) networks was also considered. The realization of the model allows one to obtain numerical information on the following questions of forward planning: nodes whose throughputs are critical and rational allocation of appropriations for their reconstruction, a rational scheme of flows of transport facilities with allowance for the minimization of operating costs and volumes of empty flows, and the determination of an optimal structure of a fleet of transport facilities and rational use of appropriations for its extension.

At an informal level, the considered problem of forward planning is described as follows. A scheduled period is given that consists of separate intervals. Each interval of the scheduled period is characterized by a rather stable freight traffic flows on a railway network. As a rule, the choice of intervals is determined by a seasonal factor. For the scheduled period, the following main parameters of capabilities of a transport system are determined: its fleet of transport facilities and cargo-handling capacities of stations. A plan of reconstruction of a transport system lies in the determination of an expedient allocation of a given amount of financial investments destined for the increase in cargo-handling capacities of stations and replenishment of the rolling stock.

According to the optimization criterion, optimization consists of the maximization of the total profit obtained during the scheduled period owing to freight traffic. As a result of calculations, it is possible to obtain numerical information on main questions of forward planning such as:

- Determination of railway stations whose throughput capacity is critical and provision of recommendations on the allocation of appropriations for their reconstruction
- Determination of a rational structure of the fleet of transport facilities and provision of recommendations on the allocation of appropriations for its replenishment
- Development of a rational scheme of railcar traffic with allowance for the minimization of operating costs and volumes of empty flows
- Rational choice of orders for transportation to places of realization

These data can be used to make decisions on promising lines of development of a transport system and to objectively substantiate critical factors of its functioning. A scheme of railcar traffic provides important information for solving the problem of determination of routing of freight traffic and train operation scheduling.

The software support developed for the solution of such problems uses decomposition schemes, which take into account block structures, together with the r-algorithm. To solve an internal subproblem, a scheme of decomposition under constraints and interior-point methods [35] are used that make it possible to take into account the block structure of the subproblem. A software implementation of the algorithm in the first problem also requires a subroutine for finding shortest paths in a directed network, and this problem can be efficiently solved with the help of the method of sequential analysis of variants.

The considered examples of works on modeling demonstratively illustrate that, even under difficult economic conditions at the beginning of the 1990s, the Institute of Cybernetics retained a powerful scientific potential that made it possible to realize large-scale projects with creating complicated optimization models, to develop computational algorithms, and to apply this support tools to the solution of complicated applied problems.

The enthusiasm inherent in young collectives and forward-looking policy of V. M. Glushkov and V. S. Mikhalevich promoted the formation of groups of scientists with strong creative potential who efficiently worked in topical lines of investigation in the field of optimization. A striking example can be the school of B. N. Pshenichnyi who became an academician of the National Academy of Sciences of Ukraine in due time. The subject matter of his works was connected with the investigation of quasi-Newtonian methods (together with Yu. M. Danilin) [145], convex analysis in Banach spaces [144], methods of linearization and nonlinear programming [141, 142], the theory of multivalued mappings, and guaranteed estimates. Most powerful results were obtained in the field of differential games [63, 146, 177, 178].

In the 1960s, at the initiative of B. N. Pshenichnyi and with the assistance of V. S. Mikhalevich, investigations connected with pursuit-evasion games including group games were initiated, which have applications in economy and technical systems, in particular, in defense systems (air-defense systems, etc.). During these investigations, problems of nondifferential optimization also arose. Main results on these subjects are contained in [63, 177].

The development of nonsmooth optimization methods created preconditions for the research and development of optimization models with random parameters that were called stochastic optimization models. There is a wide class of applied problems that can be appropriately formulated as the mentioned models. Stochastic optimization is particularly widely applied in systems of making decisions on the choice of lines of long-term development of an economy on the whole and its separate branches (power engineering, agriculture, transport, etc.) and also in ecological–economic modeling and investigation of social processes.

In investigating problems of stochastic optimization in the Institute of Cybernetics, significant attention was given to the development of numerical methods of solution of stochastic problems and to the practical application of these methods. In particular, in the 1970s, stochastic quasigradient methods for the solution of general optimization problems with nondifferential and nonconvex functions were developed at the initiative of V. S. Mikhalevich. The possibility of application of these methods to problems with nondifferential functions is rather significant for important applied problems with a fast and unforeseen behavior of objects being modeled. Stochastic quasigradient methods can be considered as a generalization of stochastic approximation methods to constrained problems and also as a development of random search methods. The following important distinctive feature of stochastic quasigradient methods should be emphasized: They do not require exact values of objective functions and constraints. This provides ample opportunities of using them for the optimization of complex systems under uncertainty.

It should be noted that subgradient methods of nondifferential optimization that were developed in the Cybernetics Institute under the guidance of V. S. Mikhalevich and N. Z. Shor exerted influence on the development of theory and practice in many directions of investigations of mathematical programming. They have gained acceptance of leading specialists of the world's scientific community in the field of optimization. The achievements of the domestic school of nondifferential optimization were also highly appreciated in the former USSR and in Ukraine. For their fruitful work on the development of numerical optimization methods and their applications, N. Z. Shor and his disciples were awarded the State Prizes of the USSR (1981) and Ukraine (1973, 1993, and 2000) in science and engineering and the V. M. Glushkov Prize (1987) and V. S. Mikhalevich Prize (1999) of the National Academy of Sciences of Ukraine.

At the present time, the active development of nondifferential optimization methods and their applications continues. As early as 1982, N. Z. Shor and V. I. Gershovich wrote in [19] that the theory of the entire class of space dilation algorithms was still a long way off perfection and that the construction of an algorithm whose practical efficiency would not concede the efficiency of the *r*-algorithm and that would be as well substantiated as the ellipsoid method seemed to be a rather realistic objective. Although more than 25 years have passed since that time, the problem of rigorous substantiation of *r*-algorithms still remains topical until now. As a step in this direction, we can mention [167] in which the transformation of a special ellipsoid into a sphere is based on a space transformation close to that used in *r*-algorithms. However, space is dilated here in the direction of the difference between two normalized subgradients. It is close to the direction of the difference between two subgradients only when the norms of the subgradients are close. If the norms of the subgradients are considerably different, then the result will be essentially distinct from the space dilation in the direction of the difference of two subgradients. A similar transformation of space is also used in new modifications of ε-subgradient space dilation algorithms developed by N. G. Zhurbenko [192, p. 29–39]. Investigations along these lines are still being actively developed.

2.3 Problems in the Development and Use of Systems Analysis Methods

Studying systems of different nature and adapting methods intended for one class of problems to other classes was always one of the main trends in the development of cybernetics. The Glushkov Institute of Cybernetics always took an active part in such studies, which were later called "systems analysis."

Systems analysis means studying an object as a set of elements that form a system. In scientific research, it is intended to study the behavior of an object as a system with all the factors that influence it. This method is widely applied in integrated scientific research of the activity of production associations and the industry as a whole, in determining the proportions of the development of economic branches, etc.

There is no unified technique for systems analysis in scientific research yet. In practical studies, it is applied in combination with operations research theory, which allows quantitative estimation of the objects of study, and analysis of systems of study of objects under uncertainty, using system engineering to design and synthesize complex systems during the study of their operation, for example, to design and estimate the economic efficiency of computer-aided control systems for technological processes. Systems analysis is important in human resource management.

As far back as in the early 1960s, Mikhalevich drew the attention of employees of the Institute of Cybernetics to recognition and identification theory for stochastic lumped- and distributed-parameter systems. The great applied importance of these studies stimulated the development of new subject areas in the statistics of random functions. New approaches to the analysis of problems in nonlinear and nonparametric regression analysis were proposed, new classes of estimates and their asymptotic properties were investigated, and stochastic optimization and estimation theories were related. In this field, noteworthy are the studies [78, 80, 81].

In a number of studies, Mikhalevich developed modern notions of informatics and its interaction with allied sciences such as cybernetics, mathematics, economy, etc.

He investigated the trends in the formation of informatics as a complex scientific discipline that studies all aspects of the development, design, creation, operation of complex systems, their application in various fields of social practice, ways of formalization of information for computer processing, and development of information culture of the society [97, 107]. By virtue of the aforesaid, informatics should widely use methods of systems analysis.

The breadth of interests of Mikhalevich was huge. In particular, by his initiative, Yu. M. Onopchuk studied regulation processes in the major functional systems of living organisms: circulatory, respiratory, and immune [137].

The original scientific work [110], which is a vivid example of the versatility of interests of Mikhalevich, was carried out jointly with V. M. Kuntsevich. It is devoted to the extremely vital (at the late 1980s) problem of identifying the causes of the arms race (primarily nuclear) between the superpowers going on at that time. The model was a discrete game with two terminal sets. Such problems are distinguished in dynamic game theory as extremely difficult; nevertheless, the use of the methods of general control theory and systems analysis allowed conducting a qualitative analysis of this phenomenon and formulating relevant recommendations. Indirectly, the authors estimated the efficiency of investment strategies for conflict interaction projects. Important studies into discrete optimization, in particular, solution of discrete–continuous problems, were conducted at the department headed by I. V. Sergienko, in close cooperation with Mikhalevich's department. In addition to creating specialized approximate local algorithms according to the descent vector method, they considered solution stability, general problems in global optimization, sequential search, multicriterion optimization, the asymptotic complexity of simple approximate algorithms for solving extremum problems on graphs (together with Prof. V. O. Perepelytsya), problems on permutations (together with Belarusian mathematicians V. A. Emelichev and V. S. Tanaev) and others. These studies were accompanied by large-scale computing experiments and development of application software.

In the 1950–1960s, the rapid development of science and technology made it necessary to use a computer to solve complex scientific and engineering problems unsolvable by traditional methods and to which systems analysis methods were applied. To solve such problems, the power of available computing facilities had to be used to the greatest possible extent. This resulted in a new scientific field, computation optimization theory, Mikhalevich being one of the founders.

Mikhalevich was actively engaged in training and professional development of specialists in optimization, systems analysis, and operations research. He was a cochairman of the majority of Winter Schools on Mathematical Programming in Drohobych and Summer Schools on Optimization, which were conducted by academician N. N. Moiseev. Scientists of the Institute of Cybernetics participated in Mathematical Schools in Sverdlovsk (organized by academician N. N. Krasovskii) and at the Baikal lake (organized by Irkutsk scientists), in All-Union Seminars on graph theory, etc.

Let us now dwell on the cooperation with the International Institute for Applied Systems Analysis. The Academy of Sciences of Ukraine placed strong emphasis on this institute, which was launched in 1972 with active participation of

Glushkov and Mikhalevich. The International Institute for Applied Systems Analysis (IIASA) is a unique nongovernmental institution engaged in scientific study of global problems that can be solved only at the international level. The number of such problems and their scope are growing, which is demonstrated by the current global crisis of financial and economic systems of different countries. It shows that the lack of systemic analysis and coordination at the international level can easily lead the global economy to a megacatastrophe. It also demonstrates the need for an independent scientific institute, common to all countries and engaged in research and solutions to global interdisciplinary problems. The issues IIASA deals with include safe, in the broadest sense, development of global power engineering based on new energy-saving and environmentally friendly technologies and related climate change, sustainable development of agriculture and food production, and water and demographic problems, including the possibility of large-scale migration of population, animals, and insects that can adversely affect the health system. All this requires new models and methods to be developed for robust (stable) solutions under uncertainty and impossible accurate forecast of various risks.

An attractive feature of IIASA in terms of scientific research is its private nature. IIASA activities are coordinated by national systems analysis committees of member countries, which (by tradition, beginning with the USA and USSR and other countries, including Ukraine) are created by national academies of sciences. Since the global issues affect the interests of various organizations, national committees include representatives of leading organizations that are directly related to specific IIASA programs.

International Institute for Applied Systems Analysis (Laxenburg, Austria)

Obviously, the search for solutions to these problems taking into account the specifics of individual countries is only possible through cooperative efforts of researchers from different countries, which is the main task of the IIASA. In this sense, the IIASA can be considered a Ukrainian Research Institute, Institute of the USA, Russia, Germany, Japan, Poland, and all other countries that are its members. Along with these countries, Austria, Egypt, India, China, the Netherlands, Norway, Pakistan, Sweden, South Africa, and South Korea are currently members of the IIASA.

The IIASA is located 16 km from the Austrian capital Vienna, in the picturesque town of Laxenburg. From the USSR, the chairman of the Council of Ministers of the USSR A. M. Kosygin and academician D. M. Gvishiani and, from the USA, President L. Johnson and his adviser M. Bundy had a great influence. Academician V. M. Glushkov and Professor H. Rayffa from Harvard University (USA), widely known for numerous monographs in optimal solutions, provided substantial assistance. Professor Rayffa was appointed the first director of the IIASA, and Gvishiani became the Chairman of the Council, which manages the IIASA.

Extensive experience in solving complex multidisciplinary problems was accumulated at the Institute of Cybernetics prior to 1972. In fact, the management of the Academy of Sciences of the UkrSSR deeply understood the need for comprehensive cybernetic or, as it is often said nowadays, systemic approaches long before the formulation of the basic ideas of the IIASA. Much of it corresponded to the Soviet approach to the coordination of various areas on the basis of interrelated planning. It was obvious that the practical implementation of these approaches to solve complex scientific problems is only possible based on computers. The first in the USSR MESM, available at the Institute of Cybernetics, attracted a wide range of researchers from different organizations in the former Soviet Union. The Institute of Cybernetics was deluged with new mathematical problems, which often arose at the junction of various sciences and could not be solved by traditional methods.

As already mentioned, the faculty of cybernetics was established at the Shevchenko State University of Kyiv to develop cybernetic approaches. The Institute of Cybernetics included a Department of Moscow Institute of Physics and Technology and a Training Center to teach high-ranking managers responsible for making governmental decisions. After the creation of the Institute of Cybernetics, cybernetic approaches based on the use of computers and mathematical models of new type became the primary activities of many organizations and educational institutions of the Soviet Union.

From the formal mathematical point of view, these problems were similar to those that arose in different divisions of the IIASA since its creation. The need for considering the diversity of system solutions often led to new optimization problems characterized by extremely high dimension and substantial uncertainty.

By the time of formation of the IIASA, the Institute of Cybernetics accumulated considerable experience in solving such problems. First of all, fundamentally new methods of nondifferential and stochastic optimization that significantly differed from the Western approaches were developed. Since the original problems that require systemic or cybernetic approaches were similar, it was decided to compare the methods being evolved and to create more efficient hybrids that can be applied to various IIASA applied problems.

This idea was proposed to the IIASA by Western scientists and was actively supported by Glushkov and Mikhalevich. While the initial cooperation of the Institute of Cybernetics and IIASA began with the development of computer networks, which provided an opportunity for our organizations to use Western computer networks, in 1976 the cooperation center moved to new models and methods for solving important problems of systems analysis. In particular, great attention was paid to the development of approaches to the analysis of robust solutions for high-dimensional optimization problems under substantial uncertainty. Two types of problems usually arise in this case. They correspond to the cases of over- and underestimated risk and result in optimization models with nonsmooth and discontinuous functions. In particular, overestimated capabilities of a system can have large-scale consequences (like the Chernobyl disaster) and involve a large number of variables for modeling. For example, the IIASA's stochastic model for the analysis of robust fields of the development of the world's power engineering had up to 50,000 variables and took into account critical nonconvex effects of increasing returns (profits) of new technologies and the uncertainty of their appearance on the world market, which required the use of powerful Cray T3E-900 supercomputer at the Computing Center of the U.S. Department of Energy.

Since 1976, Mikhalevich and his student Ermoliev started permanent cooperation with the IIASA employing leading Western experts in optimization such as G. Dantzig, R. Wets, and T. Rockafellar. Many Nobel laureates such as Arrow, Koopmans, Prigogine, Klein, Kantorovich, Crutzen, and Schelling came with short- and long-term visits.

The far-sighted policy of Glushkov and Mikhalevich directed to work closely with the IIASA had important consequences. The relations with leading Western organizations and scientists were established, which made it possible to quickly evaluate and review the experience gathered at the Institute of Cybernetics in modeling and solution of complex problems. The essential feature of the IIASA, which has a small permanent staff, is its extensive network of relationships with leading research organizations and scientists all over the world. This makes it possible to organize the world experience, find relevant information, formulate a problem adequately, and find solutions that reflect the experience of many countries, rather than one-sided recommendations of a single, even leading country.

The illustrative example that demonstrates the effectiveness of the international cooperation at the IIASA may be the project of development of models and stochastic-programming methods for the analysis of optimal strategic decisions under uncertainty. Considering the time it took well-known teams of researchers from different countries to develop models and methods, acquire and analyze data for various applied problems, and create software that appeared in the final project report, it was calculated that the project completed at the IIASA approximately in 2 years from 1982 required about 200 person-years for its implementation. From the

IIASA, only two scientists (R. Wets from the USA and Yu. Ermoliev from the Institute of Cybernetics) worked on the project. Through a network of researchers from IIASA member countries, they managed to bring together theoretical and practical developments of various scientific schools, gained practical experience and contemporary software. This was published as a joint monograph and became public. This, one of the numerous IIASA projects, also led to the international stochastic-programming community, which now includes a wide range of researchers from different countries and organizations. The community regularly holds conferences and organizes joint projects, publications, and training.

The nongovernmental status and the research orientation of the institute allow selecting candidates for vacant research positions not according to quotas but by matching their research interests and qualification to the main areas in the project where a vacancy appeared. Each applicant is interviewed by institute program managers and typically must make a scientific report. The final decision is made by the program leader who completely manages the budget allocated to the institute. Noteworthy is that the employees of the All-Union Institute of Systems Studies headed by D. M. Gvishiani, Chairman of the Council of the IIASA, worked at the IIASA only for the first years, while employees of the Institute of Cybernetics worked at the IIASA all the time.

The influence of Mikhalevich at the IIASA began to grow significantly since the mid-1980s. Previously, the status of the IIASA was shaken by the Republicans who come to power in the USA and did not support multilateral cooperation and refused to pay membership fees. In addition, Gvishiani voluntarily resigned from the position of the chairman of the IIASA Council and necessitated chairman reelection. A critical issue was the choice of a new chairman of the council, who could convince decision makers in the Soviet Union to continue paying membership fees, which equally with the former US fees made major contributions to the IIASA budget. Although Nobel laureates' letters in support of the IIASA could partially compensate for the US fee from other sources, the situation actually questioned the existence of IIASA.

In June 1987, the IIASA Council elected academician Mikhalevich as the new chairman of the council, considering that he was the director of the Institute of Cybernetics, which then had wide relationships in the USSR and successfully cooperated with almost all parts of the IIASA, clearly demonstrating the benefits of international scientific cooperation.

It should be noted that the council of the IIASA and thus the position of the chairman are very important for that institution. The main objective of the council is a strategic plan, including the selection of promising areas of research, the institute's budget allocation among its departments, admission of new and exclusion of insufficiently active member states, election of the director who was to follow the decisions of the council. The council meets twice a year; its decisions are made by voting of representatives of the member states, one representative from each state. Preparing and making decisions is a rather complicated and long-term process that requires proposals to be approved by all the member countries between meetings, the chairman playing an active role and providing permanent consultations.

Mikhalevich successfully coped with the difficult task of consolidating IIASA members with a strong financial and scientific support from the USSR. To coordinate the work with the IIASA, a new National Committee in Systems Analysis was created at the Presidium of the Academy of Sciences of the USSR, which included representatives of the relevant organizations of the Academy of Sciences and other organizations directly related to the IIASA studies. A Systems Research Laboratory was established at the Institute of Cybernetics at Ermoliev's department. Its tasks included the analysis and distribution of IIASA's results among stakeholders in the USSR in the form of short statements, analytical notes, and speeches with overview reports at various conferences.

Unfortunately, the declaration of independence caused almost unresolvable situation with the further participation of Ukraine in the IIASA because of the necessity to pay membership fees. In this situation, B. E. Paton, president of the National Academy of Sciences, and Mikhalevich applied to the council to allow Ukraine to be admitted as a member country at the IIASA as an exception, while bound to organize groups of employees at the institutions of the Academy of Sciences to perform tasks in IIASA programs equivalent to the fee of Ukraine. This letter was signed by Paton, and the council made favorable decision due to the efforts of Mikhalevich and Shpak who became a representative of Ukraine to the IIASA Council after the death of Mikhalevich. Naturally, the services of the National Academy of Sciences of Ukraine to the IIASA since its establishment were taken into account.

The work of employees of leading institutes of the National Academy of Sciences (NAS) for IIASA programs in power engineering, economy, agriculture, control of catastrophic risks, demography, and environmental control enrich their understanding of global problems and of the approaches to their analysis dominating over the world. In many respects, this prevents one-point views in Ukraine, which reflect interests of isolated countries, say, from Europe or the USA.

Noteworthy is that different models developed in the IIASA are intended to evaluate the contribution of individual countries and regions to the global situation and to analyze the prospects of joint mutually beneficial development. These models include massive bases of mutually coordinated initial data, predictive information, expert estimates, and scenarios of uncertainties, which can affect the sustainable development of systems under study. The models are refined and developed, specific formal and informal methods of their analysis, including the analysis of obtained multidimensional solutions, are accumulated. All this is very rich material for member countries (including Ukraine), which can use original IIASA models directly or create their modifications and simplified versions on their own account. Along with global power engineering models, the IIASA has a global model of manufacturing agricultural products and meal, a model of interstate (interregional) transmission of air pollution and their influence on the environment and the human, which is constantly used in negotiations about the reduction of pollution emissions by countries and individual enterprises, and a model of estimating global demographic tendencies in different countries. Leading institutes of the NAS of Ukraine and other Ukrainian organizations participate in the development of individual modules of these models and use the results in their study during the preparation of strategic documents related to the development of power engineering in Ukraine, demographic and social issues, safe development of rural territories and food stuff manufacture, prospects of the development of biofuel, trade, and greenhouse gases emission. Together with the IIASA, the V. M. Glushkov Institute of Cybernetics and a number of other institutes of the Cybernetics Center of the NAS of Ukraine continue the active development of new models and methods that adequately reflect current global changes. For example, new approaches to the management of catastrophic risks related to large territories and a great number of people are developed. In systems analysis of processes with possible catastrophic consequences, the central issue is the search for robust solutions based on stochastic optimization methods being developed at the V. M. Glushkov Institute of Cybernetics and IIASA, in scientific organizations of the USA, Norway, and Great Britain.

Thus, the participation of Ukraine in the activity of the International Institute for Applied Systems Analysis, on the one hand, bears witness to the fact that the international scientific community recognizes the high level of studies conducted by Ukrainian research institutions, primarily by those of the NAS of Ukraine, and on the other hand, it is a priority in the international cooperation of Ukraine since it opens ample opportunities for the participation of Ukrainian scientists in important scientific international projects.

Undoubtedly, the recognized authority of the Ukrainian science at the IIASA, active participation of scientists from Ukraine in programs of the institute is in many respects a great merit of academician Mikhalevich. His name is cherished at the International Institute for Applied Systems Analysis. The council of the institute established the annual Mikhalevich Award to encourage the best scientific study performed by young scientists from different countries during their participation in the IIASA Summer School. It is symbolic that many young scientists from Ukrainian research institutes take part in this Summer School.

2.4   Stochastic Optimization Methods

In the first chapter, we briefly reviewed the first research projects conducted by V. S. Mikhalevich under the supervision of B. V. Gnedenko, academician of the National Academy of Sciences of Ukraine, while he studied at the Shevchenko State University in Kyiv and under the supervision of academician A. M. Kolmogorov during the postgraduate study at the Lomonosov Moscow State University. These works were published in authoritative scientific journals and at once drew the attention of experts in the statistical decision theory, which made him one of the most cited authors in statistical analysis. Mikhalevich obtained outstanding results on sequential Bayesian solutions and optimal methods of acceptance sampling. He developed the technique of recurrent relations for risks, which made it possible to solve new problems arising in various areas of science and technology.

Mikhalevich was the first to justify Wald's problem of comparing two simple hypotheses on the mean value of a discrete-time Wiener process and to prove that a Bayesian solution can be obtained by solving the Stefan problem with moving boundaries, which is well known in the theory of differential equations. Note that in the case of discrete time, it is in many cases possible to qualitatively describe Bayesian solutions and the domains in which the observation should be terminated to accept a hypothesis. However, it is often very difficult to describe such domains, and a specific form of solution cannot always be obtained; if it can, it is very awkward. Mikhalevich proposed a method to solve the problem by passing to the limit in a discrete-time model to make it continuous time, which allows obtaining optimal solutions. As he proved, if it is possible to pass to the limit in the discrete-time model, the risk satisfies some parabolic differential equation and finds the bounds reaching which requires terminating the observation process and making some decision. This involves difficulties related to finding the optimal solution, control, and objective function for stochastic continuous-time systems. Note that the results are far beyond the scope of optimal statistical control problems and, in fact, form the basis of optimal control theory for stochastic discrete- and continuous-time systems.

Later on, the methods of reduced Bayesian strategies developed by Mikhalevich were used as the basis for solving important economic problems related to optimal planning and design.

It should be emphasized that Mikhalevich paid much attention to new fields in optimization theory, he tried, in every possible way, to encourage talented young people and unveil their creativity. Among such fields, we will dwell on stochastic optimization and identification theory, optimal control theory, and risk assessment. The problems that were studied lacked complete information on objective functions, constraint functions, and their derivatives.

A comprehensive system analysis of the future development of economy, power engineering, and agriculture and their influence on the environment and a human being requires stochastic optimization models and methods that would make explicit allowance for the probabilistic nature of the processes and the risk due to the uncertainty inherent in the decision-making process. Moreover, there is a wide class of applied problems that cannot be formulated and solved within the deterministic framework. Being unable to list all such problems, we will nevertheless mention problems related to inventory theory, maximum average network time, generalized Polya scheme with increments that have a random number of balls (these problems necessitate analyzing processes with discontinuous urn functions). These problems are widely applied in economics, biology, chemistry, etc., and are discussed in [3, 49, 53].

For illustration, let us present two examples of such problems.

Example 2.1. Consider a one-product inventory management model with one storehouse and random consumption. Given a storehouse of finite capacity, create a homogeneous stock x of the product for which demand is a random variable. The costs of storage of non-sold product and pent-up demand can be described by the function

$$f^0(x,\ \omega) = \begin{cases} \alpha(x - \omega), & \text{if } x \geq \omega, \\ \beta(\omega - x), & \text{if } x < \omega, \end{cases}$$

where α is the cost of storage of a product unit and β is pent-up demand penalty. It is obvious that $f^0(x, \varpi)$ is a convex and nonsmooth function. The task is to minimize $f^0(x, \varpi)$.

Example 2.2 (knapsack problem). The problem is formulated as follows: minimize

$$F(x) = E \sum_{i=1}^{n} d_i \ max \ [0, (\theta_i - x_i)]$$

subject to the constraints

$$\sum_{i=1}^{n} c_i x_i = a, \quad x_i \geq 0, \quad i = 1, 2, ..., n,$$

where x_i is the amount of the ith product ordered; d_i is the loss because of incomplete delivery of a unit of the ith product; θ_i is the demand for the ith product, which is a random variable with known (preset or obtained from analytically processed observation results) distribution; c_i is the cost of a unit of the ith product; and a is the product inventory.

Sometimes, this problem is subject to the constraint

$$0 \leq x_i \leq a_i, \quad i = 1, 2, ..., n,$$

where a_i is the maximum order volume depending on the capacity of the storehouse occupied by the ith product. In our case, $F(x)$ is a nonsmooth function.

Stochastic Optimization. This field of research was headed by Yu. M. Ermoliev, Academician of the National Academy of Sciences of Ukraine, a cofounder of the well-known Ukrainian school of optimization. The direct stochastic-programming methods developed by Ermoliev made him recognized worldwide, became classical, appeared in almost all tutorials on stochastic programming, and referred to in many scientific papers and monographs. The international community highly appreciated his achievements and awarded him (as one of the world-famous founders of stochastic-programming theory) a medal and an international prize for prominent services to the cause of developing the theory of financial and insurance mathematics.

Any optimization model, such as

$$\min \{f^0(x, \varpi) \mid f^i(x, \varpi) \leq 0, \ i = 1, \ldots, m; \ x \in X \subset R^n\},$$

includes parameters $\varpi \in \Omega$, which may generally be random either because of incomplete information about their values due to, for example, measurement errors, the use of their statistical estimates or because of the stochastic nature of the parameters such as weather forecasts, variations of price, demand, productivity, share prices, etc. Then, the constraint $f^i(x, \varpi) \leq 0$ can fail for any fixed $x \in X$ and some realizations of the random parameter ϖ. The set of functions,

$$f(x, \omega) = \{f^0(x, \omega), \ f^1(x, \omega), \ldots, f^m(x, \omega)\}, \quad \omega \in \Omega,$$

can be considered as a vector characteristic of a solution $x \in X$, and finding the optimal solution can generally be regarded as a vector optimization problem with an infinite number of criteria.

Optimization problems with random parameters are formalized in stochastic-programming theory. The stochastic quasigradient methods proposed by Ermoliev underlie direct stochastic-programming methods and can be considered a generalization of stochastic approximation methods on a constrained problem, the Monte Carlo method on an optimization problem, and a development of random search methods. The key feature of quasigradient methods is that they use statistical estimates (rather than exact values) of objective functions and constraints obtained from realizations of subintegral functions and their generalized gradients. This opens ample opportunities for these methods to be applied to optimize complex stochastic systems by simulation modeling.

Among the pioneering studies that underlie stochastic quasigradient methods, noteworthy are [53, 59, 62] and some others written in the late 1960s and early 1970s and discussing the solutions of both nonlinear and convex stochastic-programming problems with general constraints. It should be underlined that objective functions and constraints that are not necessarily differentiable were considered. The general idea of stochastic quasigradient methods is as follows.

Let us minimize a convex objective function $F^0(x)$ subject to the convex constraints $F^i(x) \leq 0$, $i = 1, \ldots, m$; $x \in X \subset R^n$, where the functions may have the form of expectations. When deterministic optimization methods are used, the sequence of approximate solutions $\{x^k, k = 0, 1, \ldots\}$ that minimizes the function $F^0(x)$ is usually found from the exact values of the functions $F^i(x)$, $i = 0, \ldots, m$, and their gradients (generalized gradients if the functions are nonsmooth). Nevertheless, in stochastic quasigradient methods, such a sequence is constructed based on statistically biased and unbiased estimates of the functions $F^i(x)$ and their generalized gradients $F_x^i(x)$:

$$\eta^i(k) = F^i(x^k) + a_i(k),$$

$$\xi^i(k) = F_x^i(x^k) + b_i(k), \quad i = 0, \ldots, m,$$

where the errors $a_i(k)$, $b_i(k)$, $i = 0, \ldots, m$, can depend on the path $\{x^0, \ldots, x^k\}$ for the current iteration k and tend to zero in some probabilistic sense as $k \to \infty$. The random vector $\xi^i(k)$ is called stochastic quasigradient of the function $F^i(x)$ at the point x^k for $b_i(k) \neq 0$. If $b_i(k) = 0$, then $\xi^i(k)$ is called the stochastic gradient or stochastic generalized gradient (stochastic subgradient) of the function $F^i(x)$ depending on whether it is smooth or undifferentiable. The main way of constructing stochastic generalized gradients is to introduce the differentiation or generalized-differentiation sign under the expectation sign. Assume that

$$F^i(x) = Eg^i(x, \varpi) = \int g^i(x, \varpi)P(d\varpi).$$

Then the continuously differentiable function $F^i(x)$ has the following gradient:

$$F^i_x(x) = Eg^i_x(x, \varpi) = \int g^i_x(x, \varpi)P(d\varpi),$$

and thus, the random vector $\xi^i(x, \varpi) = g^i_x(x, \varpi)$ is the stochastic generalized gradient of the function $F^i(x)$ at the point x. If $F^i(x)$ and $g^i(x, \varpi)$ are nonsmooth functions, then the stochastic generalized gradient $\xi^i(x, \varpi)$ should be taken as an intersection (measurable in a set of variables) of the subdifferential $g^i_x(x, \varpi)$ and subintegral function $g^i_x(x, \varpi)$. Another way to construct stochastic quasigradients is based on the finite-difference method. For example [27], the stochastic quasigradient of the function $F^i(x)$ at the point x^k can be calculated by the formula

$$\xi^i(k) = (3/2)[g^i(x^k + \delta_k h_k, \varpi^k) - g^i(x^k, \varpi^k)]h^k/\delta_k,$$

where h^k is independent observation of the random vector $h = (h_1, \ldots, h_n)$ with independent components uniformly distributed over the interval $[-1, 1]$. Note that to find the vector $\xi^i(k)$ in this case, it is necessary to calculate the values of the function $g^i(x, \varpi)$ only at two points, irrespective of the space dimension n. If the functions $g^i(x, \varpi)$ have complex implicit structure and are defined, for example, on the solutions of differential equations (which is typical of ecological applications) or by simulation models, then the use of the random finite-difference direction $\xi^i(k)$ is advantageous over the standard finite-difference approximation of the gradient.

To solve the problem

$$\min_{x \in X}[F^0(x) = Eg^0(x, \varpi)] = F^*,$$

the stochastic quasigradient method suggests constructing a sequence of approximations $\{x^k, k = 0, 1, \ldots\}$ as follows [53]:

$$x^{k+1} = \Pi_X\{x^k - \rho_k \xi^0(k)\}, \ k = 0, \ 1, \ldots,$$

where $\Pi_X\{\cdot\}$ is the operator of projecting onto the convex compact set X,

$$E\{\xi^0(k) \mid x^0, \ldots, x^k\} \in F_x^0(x), \ E\|\xi^0(k)\|^2 \le \text{const},$$

the step factors satisfying the following conditions with probability one:

$$\rho_k \ge 0, \ \sum_k E\rho_k = +\infty, \ \sum_k E\rho_k^2 < \infty.$$

Under these assumptions, the sequence $\{x^k\}$ coincides with the optimal solution x^* of the problem. The optimal value F^* can be estimated from the independent observations $f^0(x^k, \omega^k)$ of values of the subintegral function by the formula

$$F^k = (1/(k+1)) \sum_0^k f^0(x^i, \omega^i) \to F^* \text{ as } k \to \infty.$$

Let us dwell on one possible application of stochastic quasigradients where stochastic decomposition methods are used to solve multistage problems. The basic two-stage stochastic-programming model can be formulated as follows. Two stages of planning can be distinguished: the current state (first stage) and the future state (second stage). A decision $x \in R^n$ at the current instant of time is made under inexact information on the future state w. For example, the next year's production volume x is planned under unknown demand and prices.

The task is to choose a long-term plan x stable in some sense against possible variations of unknown parameters w (demand, prices, resources, weather conditions, possible sociopolitical situations, etc.). The concept of stability of the solution x is associated with its possible correction during the observation of a specific state w. To this end, correction mechanisms $y(x, w)$ that optimally react to the situation w for any given x are assumed. For example, for a random demand, it is assumed that absent products can be purchased to form the inventory. In hazardous production, there should be services to mitigate the consequences of possible emergencies. Then, a two-stage stochastic-programming problem can be represented as the following linear-programming problem: minimize the function

$$cx + \sum_{k=1}^N p_k d_k y_k \tag{2.27}$$

subject to the constraints

$$A_k x + D_k y_k = b_k, \ k = 1, \ 2, \ldots, N,$$
$$x \ge 0, \ y_k \ge 0, \ k = 1, \ 2, \ldots, N, \tag{2.28}$$

where x and y_k are vectors of variables; $c, d_k, b_k, A_k,$ and D_k are vectors and matrices of corresponding dimensions; and p_k are numbers such that $p_k \geq 0,$ $\sum_{k=1}^{N} p_k = 1$. To select such numbers (weight coefficients), many methods can be used to determine, in each specific case, the stochastic properties of the decomposition procedure. For example, in global models of interacting countries, the matrix D_k describes the production capacity of an individual country, and the matrix A_k defines its relationships with the outside world (common resources, migratory processes, environmental standards). The number N can be very large. For example, if only the components of the vector of right-hand sides in the constraints of the second stage take two values independently, then $N = 2^m$, where m is the number of such constraints. Therefore, if the number m of constraints is relatively small, standard linear-programming methods become inapplicable. The stochastic approach to decomposing problems (2.27) and (2.28) into independent subproblems is as follows.

Fix $x \geq 0$ and let $y_k(x) \geq 0$ be a solution of the kth subproblem that has the following form: min $\{d_k y_k / D_k y_k = b_k - A_k x, \, y_k \geq 0\}$. Denote by $u_k(x)$ the dual variables that correspond to $y_k(x)$. Let x^s be the current value of x obtained after the sth iteration. In accordance with the probabilities $p_1, \ldots, p_N$, select the block (subproblem) k_s and determine $u_k(x^s)$. In this case, the vector $\xi(s) = c - u_{k_s}(x^s) A_{k_s}$ is the stochastic generalized gradient of the function $F(x) = cx + \sum_{k=1}^{N} p_k d_k y_k(x)$.

Such an approach makes it possible to develop easily implementable stochastic decomposition procedures stable against random perturbations. They can be applied to optimize complex stochastic systems. These procedures do not need all the N subproblems to be solved simultaneously at each iteration followed by complicated coordination of connecting variables x, which is due to the use of traditional deterministic approaches.

Formally, an elementary two-stage programming model is reduced to minimizing the function

$$F^0(x) = \mathbb{E} f^0(x, \, w), \, x \geq 0,$$

where $f^0(x, \, w) = cx + py(x, \, w) = cx + \min \{ py / Dy = b - Ax, \, y \geq 0\}$. Note that all the coefficients $w = \{p, \, b, \, A, \, D\}$ can be random variables. The equations $Ax + Dy = b$, where A and D are some matrices, $x \in R^n$, $y \in R^r$, and $b \in R^m$, reflect the stochastic balances between cost and production output in the system being modeled. The quantity py determines the cost of correcting the above-mentioned balance using the vector y. The objective function $F^0(x)$ of the model is convex but may generally be nondifferentiable since there is minimization operation under the expectation sign in the definition of $F^0(x)$. The values of the function $F^0(x)$ can be found analytically only in exceptional cases. If the state w has a finite number of possible values, then the model formulated has a block structure, as discussed above.

The stochastic subgradient $\xi^0(s)$ of the objective function of a two-stage problem can be written as $\xi^0(s) = c + u(x^s, w^s) A(w^s)$, where w^s, $s = 0, 1, \ldots,$ are independent observations of random parameters w; $u(x^s, w^s)$ are dual variables that correspond to the optimal plan $y(x^s, w^s)$; x^s is the current approximation to the optimal solution; and $A(w^s)$ is a random realization of the matrix A that corresponds to the observation w^s. Thus, the use of a random vector $\xi^0(s)$ and one of the versions of the stochastic quasigradient procedure allows us to avoid virtually insuperable difficulties in evaluating multidimensional integrals (expectations).

Later on, Ermoliev assembled a team of graduates of the Mechanics and Mathematics Faculty and the Faculty of Cybernetics of the Taras Shevchenko National University of Kyiv and Moscow Institute of Physics and Technology. They did fundamental research in stochastic-programming theory, which was recognized worldwide. One of the main problems in stochastic optimization theory is to extend the concept of stochastic gradient to wider classes of stochastic extremum problems. The results obtained by Yu. M. Ermoliev, A. A. Gaivoronskii, A. M. Gupal, A. N. Golodnikov, V. I. Norkin, S. P. Uryas'ev, and others considerably extended the practical capabilities of direct stochastic-programming methods. Let us briefly discuss some of them. The stochastic quasigradient method was generalized to nonconvex stochastic-programming problems with weakly convex [131], generalized differentiable [60], almost differentiable [106], and locally Lipschitz [27] functions. The modifications of the method related to averaging of stochastic generalized gradients [27], solution of nonstationary stochastic problems [18, 131], analysis of limiting extreme stochastic-programming problems [18], and adaptive step control [174] were developed and analyzed. The concept of stochastic quasigradient was used in [48, 53, 106] to develop stochastic analogues of the following constrained optimization methods: methods of penalty functions, linearization, reduced gradient, Lagrangian multipliers, methods for solving stochastic minimax problems, etc. Mikhalevich analyzed random search procedures for solving optimization problems formulated in terms of binary preference relations. These relations are a mathematical model for the pairwise comparison of alternatives in interactive decision-making algorithms. The interpretation of such procedures as a special case of stochastic quasigradient methods made it possible to propose algorithms stable against random errors of decision makers. Due to the wide use of cluster computers, methods of parallel stochastic optimization developed in [50] were widely applied. It should be emphasized that stochastic quasigradient methods were widely used to solve general optimization problems with nondifferentiable and nonconvex functions. The possibility of applying these methods to problems with nondifferentiable functions is very important for the solution of applied problems with fast and unpredictable behavior of objects being modeled, for the optimization of complex systems under uncertainty. The scope of application was also substantially extended after the analysis of non-Markov (with dependent observations) stochastic optimization procedures whose convergence in some probabilistic sense was proved [61, 75].

The other series of studies was concerned with the asymptotic properties of stochastic quasigradient methods (limiting theorems, convergence rate) and confidence intervals for iterative stochastic optimization procedures. The results were reported in the monograph [75].

Of importance is the development of stochastic discrete optimization methods. To solve stochastic global and stochastic discrete optimization problems, Yu. M. Ermoliev, V. I. Norkin, B. O. Onishchenko, G. Pflug, A. Rushchinskii, and others developed the stochastic branch-and-bound algorithm.

To estimate the optimal values, use was made of the permuted relaxation of stochastic-programming problems, which implies permutation of the operations of minimization and expectation (or probability):

$$F^*(x) = \min_{x \in X} E f^0(x, \varpi) \geq E \min_{x \in X} f^0(x, \varpi) = F_*(x),$$

$$P_* = \max_{x \in X} P\{f(x, \varpi) \in B\} \leq P\{\exists x(\varpi) \in X : f(x'(\varpi), \varpi) \in B\} = P^*(x).$$

Thus, to obtain the lower-bound estimate $F_*(X)$ of the optimal value $F^*(X)$, it is sufficient to efficiently solve the problem $\min_{x \in X} f(x, \theta)$. To obtain the upper-bound estimate $P^*(X)$ of $P(X)$, it is sufficient to test the conditions $f(x, \theta) \leq 0$ for compatibility for fixed θ, which is possible for many applications (see [198] for some of them).

It should be emphasized that along with direct stochastic optimization, other stochastic-programming methods are also developed; they are outlined, for example, in the monographs [198, 211, 212]. One of such methods is the so-called method of empirical means. It approximates the expectation functions $F^i(x) = E g^i (x, \varpi)$ by their empirical estimates

$$F^i(x, N) = (1/N) \sum_{0}^{N} g^i\left(x, \varpi^k\right)$$

and passes to approximate deterministic optimization problems of the form

$$\min \{F^0(x, N) | \, F^i(x, N) \leq 0, \, i = 1, \ldots, \, m; \, x \in X \subset R^n\},$$

to which well-known deterministic optimization methods can then be applied. Various aspects of the convergence of the method of empirical means (almost sure convergence with respect to functional or solution, convergence rate, and asymptotic convergence principle) were examined in the monograph [206] and in papers by Yu. M. Ermoliev, P. S. Knopov, V. I. Norkin, and M. A. Kaizer. The traditional and best-known approach to the analysis of this method was elaborated by R. Wets and his colleagues, and was widely developed in studying many classes of stochastic-programming problems. However, there is a wide range of problems that need new, alternative approaches to the analysis of the method of empirical means to be developed. In particular, stochastic optimization methods are widely applied to solve identification problems [54], where it is important to analyze the behavior of solutions with probability one. New approaches are widely used in modern asymptotic estimation theory. It is the method of empirical means that

allows deeper understanding of how the stochastic optimization theory is related
to the statistical estimation theory. It made it possible to describe all the main
classes of statistical estimates and, moreover, to study new classes of robust
estimates if there is a priori information on the unknown parameters and
observations are dependent random variables. This, in turn, allowed developing
new nonlinear and nonparametric methods to estimate the parameters of stochastic systems based on incomplete observations and a priori constraints for
unknown parameters and studying their asymptotic properties. Such methods
are often used in the theory of estimating parameters of random variables,
processes, and fields with special (risk) functionals, in classification and recognition problems. The asymptotic properties of estimates and the general conditions
for the convergence (in some probabilistic sense) of an approximation problem to
the original one were established. This problem was also studied in the case of
dependent observations and in the case where the unknown parameters were
elements of some functional space. The results are outlined, for example, in
[78, 206].

Along with establishing the convergence conditions for estimates, it is important
to analyze their asymptotic distribution. A priori constraints for unknown
parameters may cause estimates not to be asymptotically normal. The key result
is that finding the limiting distribution reduces to some quadratic-programming
problem [206]. Estimating the convergence rate of the method of empirical means
is also very important. Theorems on large deviations of the difference in certain
measure between an approximate value of an optimum point and its exact value
were proved.

Stochastic Optimization and Robust Bayesian Estimates. The studies in this field
involved finding Bayesian estimates under incomplete information on their a priori
distribution that, however, is known to belong to some class of functions. Such a
situation arises, for example, in estimating the unknown reliability parameters of
high-technology systems where failure statistics for system elements is very poor,
which makes it impossible to adequately choose a priori distribution. Under these
conditions, calculating the Bayesian estimates becomes a nontrivial problem.

The idea of the so-called minimax approach is to find an estimate that would
minimize the supremum of the objective functional with respect to a given class Γ
of a priori distribution functions.

A. N. Golodnikov, P. S. Knopov, P. Pardalos, V. A. Pepelyaev, and S. P.
Uryas'ev (see, e.g., [207]) studied the choice of the following types of objective
functionals:

1. Bayesian risk

$$r_H(\delta(x)) = \int_{\Theta} \int_{X} L(\theta,\ \delta(x)) f(x|\theta) dx dH(\theta).$$

2. A posteriori risk

$$\phi_H(\delta(x)) = \int_\Theta L(\theta, \ \delta(x)) dG(\theta|x) = \frac{\int_\Theta L(\theta, \ \delta) f(x|\theta) dH(\theta)}{\int_\Theta f(x|\theta) dH(\theta)}.$$

3. A posteriori expectation of the parameter θ for fixed sampled data x

$$\widehat{\theta}_H = \frac{\int_\Theta \theta f(x|\theta) dH(\theta)}{\int_\Theta f(x|\theta) dH(\theta)}.$$

   To analyze the sensitivity of Bayesian estimates to the choice of the a priori distribution function from the class Γ, it is necessary to find the lower and upper bounds of the range of possible values of the objective functional. According to the Bayesian approach, if $H(\theta)$ is a "true" a priori distribution function, the quality of the Bayesian point estimate $\widehat{\theta}_H(x)$ is measured in terms of the Bayesian risk $r_H(\widehat{\theta}_H$

$(x))$ or a posteriori risk $\phi_H(\widehat{\theta}_H(x))$. Let not one a priori distribution function $H(\theta)$ but rather a class Γ of such distribution functions be available. In this case, the values of $r_H(\widehat{\theta}_H(x))$ or $\phi_H(\widehat{\theta}_H(x))$ can no longer be used to characterize the quality of the Bayesian point estimate $\widehat{\theta}_H(x)$. The range $(r_*(\widehat{\theta}_H(x)), \ r^*(\widehat{\theta}_H(x)))$ of possible values of the Bayesian risk or the range $(\phi_*(\widehat{\theta}_H(x)), \ \phi^*(\widehat{\theta}_H(x)))$ of possible values of a posteriori risk, which are defined for all the a priori distribution functions from the class Γ, are more adequate for this purpose.
   For example, in estimating the reliability parameters for probabilistic safety analysis, it is important to know how wide the range $(\theta_*, \ \theta^*)$ of possible values of the a posteriori expectation of the parameter θ is for fixed sample data x obtained for any a priori distribution function from the class Γ.
   To find the lower $r_*(\widehat{\theta}_H(x))$ and upper $r^*(\widehat{\theta}_H(x))$ bounds of the range $(r_*(\widehat{\theta}_H(x)),$ $r^*(\widehat{\theta}_H(x)))$ of possible values of the Bayesian risk for the estimate $\widehat{\theta}_H(x)$ obtained on the assumption that each distribution function $G(\theta) \in \Gamma$ is a true a priori distribution function, it is necessary to solve the following optimization problems in the space of distribution functions:

$$r_*(\widehat{\theta}_H(x)) = \inf_{G \in \Gamma} \int_\Theta \int_X L(\theta, \ \widehat{\theta}_H(x))) f(x|\theta) dx \ dG(\theta),$$

$$r^*(\widehat{\theta}_H(x)) = \sup_{G \in \Gamma} \int_\Theta \int_X L(\theta, \ \widehat{\theta}_H(x))) f(x|\theta) dx \ dG(\theta).$$

They are stochastic-programming problems, where the optimization is in the space of distribution functions and the objective functionals are linear in the distribution functions.

To find the lower $\phi_*(\widehat{\theta}_H(x))$ and upper $\phi^*(\widehat{\theta}_H(x))$ bounds of the range $(\phi_*(\widehat{\theta}_H(x))$, $\phi^*(\widehat{\theta}_H(x))$ of possible values of the a posteriori Bayesian risk for the estimate $\widehat{\theta}_H$ obtained on the assumption that each distribution function $G(\theta) \in \Gamma$ is a true a priori distribution function, it is necessary to solve the following optimization problems in the space of distribution functions:

$$\phi_*(\widehat{\theta}_H(x)) = \inf_{G \in \Gamma} \frac{\int_{\Theta} L(\theta, \quad \theta_{\widehat{H}}(x)) f(x|\theta) dG(\theta)}{\int_{\Theta} f(x|\theta) dG(\theta)},$$

$$\phi^*(\widehat{\theta}_H(x)) = \sup_{G \in \Gamma} \frac{\int_{\Theta} L(\theta, \quad \theta_{\widehat{H}}(x)) f(x|\theta) dG(\theta)}{\int_{\Theta} f(x|\theta) dG(\theta)}.$$

To find the lower θ_* and upper θ^* bounds of the range $(\theta_*, \; \theta^*)$ of possible values of the a posteriori expectation of the parameter θ for fixed sample data x obtained on the assumption that each distribution function $G(\theta) \in \Gamma$ is a true a priori distribution function, it is necessary to solve the following optimization problems in the space of distribution functions:

$$\theta_* = \inf_{G \in \Gamma} \frac{\int_{\Theta} \theta f(x|\theta) dG(\theta)}{\int_{\Theta} f(x|\theta) dG(\theta)}, \quad \theta^* = \sup_{G \in \Gamma} \frac{\int_{\Theta} \theta f(x|\theta) dG(\theta)}{\int_{\Theta} f(x|\theta) dG(\theta)}.$$

The above-mentioned problems are stochastic-programming problems where optimization is in the space of distribution functions and the objective functionals are linear in the distribution functions.

The methods to solve these rather complex stochastic optimization problems in functional spaces and algorithms of their numerical implementation were developed and computational experiments were conducted. The results for classes of distribution functions with bounded moments or given quantiles turned out to be of special interest. For these cases, a stochastic optimization problem was proved to be reducible to a linear-programming problem. Robust Bayesian estimates, the lower and upper bounds for the estimates, which are a measure of indeterminacy due to a limited a priori information were found. The proposed methods were shown to substantially improve the quality of these estimates in estimating reliability parameters.

Some Problems of Controlled Discrete Stochastic Systems. The fundamental studies by V. S. Mikhalevich pioneered the research of control of random processes at the Institute of Cybernetics. They initiated a new area in the analysis of controlled random processes based on the generalization of the well-known Bayesian methods of stochastic decision theory. It is these studies that stimulated the development of various areas in the theory of controlled random processes and fields. The general theory of controlled Markov processes with discrete time crystallized in Wald's studies on sequential analysis and was then developed by R. Bellman, R. Howard, A. N. Shiryaev, A. V. Skorokhod, and others. Scientists from the V. M. Glushkov Institute of Cybernetics also greatly contributed to the theory of controlled random processes.

Let us formulate the control problem for stochastic discrete-time systems.

Let X and A be separable spaces, and Φ and Λ be the σ-algebras of Borel subsets of X and A. Let there be given a mapping F (multiple-valued function) that associates each $x \in X$ with a nonempty closed set $A_x \subset A$ so that the set $\Delta = \{x \in X, \ a \in A_x\}$ is Borel measurable in $X \times A$. The random evolution of the system is controlled by the set of transition probabilities $P\{B/x_n, a_n\} = P\{X_{n+1} \in B/X_0 = x_0, \ D_0 = a_0, \ ..., \ X_n = x_n, \ D_n = a_n\}$, where $B \in \Phi$; $(x_k, \ a_k) \in \Delta$; X_k is the state of the system at the time k, D_k is the decision made at the time k, and A_k is the control chosen at the time k, $k \leq n$.

Denote by $r(x, \ a)$ the expected losses over one period if the system is in the state x at the beginning of the period and a decision $a \in A_x$ is made. The function $r(x, \ a)$ is assumed to be bounded and measurable on Δ, $|r(x, \ a)| \leq C < \infty$, $(x, \ a) \in \Delta$.

The general feasible system control strategy is a sequence $\delta = \{\delta_1, .., \delta_n...\}$ such that the probabilistic measure $\delta_n(\cdot/x_0, ..., x_n)$ on $(A, \ \Lambda)$ is concentrated on A_{x_n}. A strategy is called stationary Markov if $\delta_n(\cdot/x_0, ..., x_n) = \delta(\cdot/x_n)$, $n = 0, \ 1, \ ....$ A stationary Markov strategy is called deterministic if the measure $\delta(\cdot/x_n)$ is concentrated at one point for any x.

Denote the class of all feasible strategies by R and the class of stationary Markov deterministic strategies by R_1 and consider two strategy optimality criteria:

1. Average cost for the chosen strategy δ

$$\varphi(x, \ \delta) = \lim_{n \to \infty} \sup \frac{1}{n+1} \mathrm{E}_x^\delta \sum_{k=0}^{n} r(x_k, \ D_k).$$

2. Expected (discounted) total cost with $\beta \in (0, \ 1)$ and strategy δ

$$\Psi_\beta(x, \ \delta) = \mathrm{E}_x^\delta \sum_{k=0}^{n} \beta^k r(x_k, \ D_k),$$

where E_x^δ is the conditional expectation corresponding to the process following the strategy δ under the condition $x_0 = x$.

A strategy δ^* is assumed optimal (φ-optimal or Ψ_β-optimal) with respect to these criteria if $\varphi(x, \ \delta^*) = \inf_{\delta \in R} \varphi(x, \ \delta)$ or $\Psi_\beta(x, \ \delta^*) = \inf_{\delta \in R} \Psi_\beta(x, \ \delta)$, respectively. The existence and uniqueness theorems for φ and Ψ_β strategies in R_1 were proved.

Let us consider an example of applying the optimal stochastic control theory to a modern inventory problem. Consider an inventory control system for one type of product, which can be replenished in continuous units. The maximum storehouse capacity is Q; thus, the inventory level takes values from the interval $[0, Q]$. The inventory level is checked periodically at discrete instants of time n, and a decision is then made to order an additional amount of goods. If the inventory level is $X_n = x \in [0, Q]$ at time $n \in N$, amount $D_n \in A_x = [0, Q - x]$ is ordered. The order is delivered instantaneously with probability $p \in (0, 1]$ and is lost with probability $1 - p \in [0, 1)$.

Let $\eta = (\eta_n : n \in N)$ be a Bernoulli sequence, where $\eta_n = 1$ means that the order made at the time n has been delivered. We assume that η_n does not depend on the history of the system until the time n inclusive.

There is a random demand ξ_n at the time n; $\xi = (\xi_n : n \in N)$ is a sequence of equally distributed independent random variables with continuous distribution $G(x)$, $x \geq 0$. These quantities are assumed to be independent of the history of the system until the time n inclusive and, moreover, $G(Q) < 1$.

The demand at the time n is satisfied by the quantity $X_n + D_n\eta_n$ at the end of the time interval $[n, n + 1)$ if possible. In the case of full or partial shortage of the goods, orders are not reserved but lost. The evolution equation for the inventory level has the form

$$X_{n+1} = (X_n + D_n\eta_n - \xi_n)_+, \ n \in N,$$

where $(a)_+ = max\ \{a, 0\}$ is the positive part of $a \in R$.

The assumptions guarantee that for the above Markov strategy, the sequence $X = (X_n : n \in N)$ is a homogeneous Markov chain for which the transition probabilities can easily be determined. Thus, if we assume that all the random variables for the model are defined on the common basic probabilistic space $(\Omega, \Im, P)$, we have a controlled random process with discrete time, phase space $X = [0, Q]$, and control space $A = \{d_a, a \in [0, Q]\}$, where d_a is a decision to order goods in amount a. In the state x, the set of feasible controls takes the form $A_x = \{d_a, a \in [0, Q - x]\}$.

Let the function $c(x)$ related to inventory supply be linearly dependent on the amount of the order to be delivered, and the loss function $f(x)$ related to inventory storage and deficiency be convex.

As was proved, φ-optimal and Ψ_β-optimal deterministic strategies exist for the model; moreover, the conditions are established whereby the optimal strategies have two-level structure, that is, there exists a point $x^* \in [0, Q]$ such that the optimal strategy has the form

$$\delta^* = \begin{cases} d_{Q-x}, & x \leq x^*, \\ d_0, & x > x^*. \end{cases}$$

Such inventory strategies are called (s, S)-strategies.

Similar strategies will also be optimal for some other models from queuing, reliability, and inventory theories. These results were obtained by P. S. Knopov, V. A. Pepelyaev, and their colleagues. Thus, finding optimal strategies reduces to some optimization problem; to this end, well-known methods can be applied. The scientists of the Institute of Cybernetics pay much attention to the development of numerical methods to find optimal strategies for stochastic dynamic systems. The monograph [55] outlines the finite-difference methods for deterministic and stochastic lumped- and distributed-parameter systems.

Currently, the theory of stochastic systems with local interaction is widely used to solve complex biological, engineering, physical, and economic problems. An example of such systems is Gibbs random fields and Ising fields, which are widely used to solve many problems in statistical physics. Under some natural conditions, these fields are Markov, which makes them useful for the description and solution of real-world problems in order to model them and make optimal decisions. Creating the optimal control theory for stochastic systems with local interaction, developing methods to estimate unknown parameters, and finding optimal solutions during their operation are important tasks. The monograph [217] outlines the fundamentals of this theory. It has numerous applications in recognition theory, communication network theory, economics, sociology, etc. For example, a Markov model that describes the behavior of companies with competing technologies is analyzed in [217]. It is required to develop a recognition strategy that makes a clever (in a sense) decision based on known observation. One of the best-known approaches in recognition theory is that based on modeling the behavior of unknown objects using Gibbs random fields, which are Markov fields with distribution function of known form dependent on unknown parameters. A recognition problem reduces to estimating these parameters. One more example of applying Markov random fields will be given below to describe catastrophe risk models.

Methods of Catastrophe Risk Insurance. The insurance of catastrophe risks is closely related to the previous subject. The question is not only technogenic and ecological catastrophes but also the insurance of events related to financial turmoil. The number of natural disasters has considerably increased for the last two decades, and relevant economic losses have increased fivefold. A series of interesting and important studies in this field is outlined, for example, in [56, 161].

As an example, we will discuss the activity of insurance companies intended to earn the maximum profit under a priori given constraints. The real expenses of such companies are uncertain. Actually, the cost of insurance contracts remains unknown for a long time, the total cost of insurance contracts or a portfolio not being additive.

As a rule, traditional insurance deals with independent risks of comparatively high frequency and small losses, which allows using the law of large numbers to formulate individual insurance strategies. The situation will drastically change if there are accidents with significant human and financial losses. Insurance companies usually allow for possible risks, each being characterized by the distribution of damage, which is mostly expressed in monetary terms. In other words, each risk is associated with possible losses w_i, with w_i and w_j, $i \neq j$, being interdependent random variables.

For example, an earthquake causes damage on a wide area, destructing buildings and roads, causing fire, etc. The risk portfolio is characterized by a solution vector $x = (x_1, \ldots, x_n)$, and its total value is defined by

$$c(x, \ w) = \sum_{i=1}^{n} c_i(x_i, \ w_i),$$

where the random insurance function $c_i(x_i, \ w_i)$ is generally nonconvex and nonsmooth with respect to the solution vector c.

Note that the central limit theorem is true for $c(x, \ w)$ under rather general conditions in the case of independent values. This allows using approaches that are based on the method of empirical means. Nevertheless, as indicated above, catastrophes cause dependent risks $w_1, \ldots, w_n$, and there is no certainty that summing risks will result in normal distribution. One more feature of such problems is the excessive sensitivity of the constraints appearing in the optimization problem to the joining of distribution. This may result in discontinuous functions and make it impossible to apply statistical approximations based on sample averages.

Noteworthy is another feature of catastrophe risk portfolio construction problems. The dependence of damage on the space and time coordinates is of crucial importance in many cases. The concurrence of two rare events in time and at a certain point has extremely small probability. The explicit form of insurance portfolio in the dynamic case can be obtained by introducing appropriate variables.

The function $c(t, x, w)$ can be introduced by the formula

$$c(t, x, w) = \sum_{i=1}^{n} \sum_{k=1}^{N_i(t)} c_i(t_i, \ x_i, \ w_{ik}),$$

where $N_i(t)$ is a random point process, w_{ik} is the realization of the random variables w_i of ith-type damage, $x = (x_1, \ldots, x_n)$ is the full vector of variables of the problem with values in some set X.

If we denote by $R(T, x)$ the profit of an insurance company and by $\{0, T\}$ the time horizon of insurance business planning, then to find the optimal insurance contract portfolio, the insurant should solve the optimization problem

$$F_T(x) = ER(T, \ x)I_{\{\tau(x) \ > \ T\}} \rightarrow \max_{x \in X}$$

subject to the constraints imposed on the probability of ruin in time T

$$\Psi_T(x) = P\{v(x) \leq T\} \leq \alpha,$$

where $\tau(x)$ is a stopping time such that $\min R(T, \ x) \geq 0$ and $R(v(x), \ x) \leq 0$; I_A is the indicator function of the set A.

The functions $F_T(x)$ and $\Psi_T(x)$ can be discontinuous for continuous distributions of the functions $R(T, x)$, which makes traditional stochastic-programming methods unacceptable to solve the problem posed. Thus, new methods are necessary. Some of them are outlined in [27] and those for insurance problems in [56].

One of such methods consists in stochastic smoothing of a risk process by replacing the original function with a sufficiently smooth approximation.

Modeling of accidents was discussed in [217].

Some New Approaches to Estimating Financial Risks. Risk management is one of the key problems in asset allocation by banks, insurance and investment companies, and other risk assessment financial institutions.

Credit risk is due to a trading partner who defaults on his duties. There are many approaches to the assessment of such risk. The best-known approach is based on estimating so-called Value-at-Risk (VaR for short), which has become a standard of risk assessment and management. VaR is known to be defined as the d-quantile of an investment portfolio loss function. Constructing a portfolio with predetermined constraints imposed on VaR or with minimum admissible VaR is an important problem.

In his Dr.Sci. thesis, V. S. Kirilyuk developed a new approach to risk assessment. It consists in estimating conditional average loss CVaR, which exceeds α-VaR. In many cases, CVaR appeared better than VaR. Though CVaR has not become a standard risk measure in the financial industry yet, it plays an important role in financial and insurance mathematics.

2.5   Discrete Optimization Methods

An important field of studies focused on the development of scientific foundation in modern computer technologies [163] is the development of optimization (in particular, discrete optimization) models and methods [17, 29, 105, 109, 116, 124, 153, 156, 158, 159]. As of today, mathematical models of discrete optimization, which has evolved for more than 40 years, cover a wide range of applied problems related to the efficient solution of numerous problems of the analysis and optimization of decision making in various scientific and practical fields. It is thus of importance to develop a mathematical theory, to improve the available mathematical models and methods as well as software and algorithms, and to develop new ones to solve complex discrete optimization problems. Nevertheless, the majority of discrete optimization problems are usually universal (NP-hard), which makes it hardly possible to create linear and convex optimization methods for them, which are as efficient as the available ones. The computational difficulties that arise in solving discrete optimization problems are often due to their high dimension, multiextremality, specific structure, multicriteriality, weak structure, and incomplete and uncertain input information on their parameters. This obviously makes it impossible to develop acceptable exact methods for most classes of discrete problems. Moreover, mathematical models (including discrete ones) of applied problems of finding best solutions do not usually represent real situations but rather approximate them. Therefore, it is expedient to solve discrete optimization problems with appropriate degree of approximation to the optimum taking into account the possible inaccuracy and incorrectness of the mathematical models. It is thus important to develop and analyze various approximate methods for solving these problems (these are usually methods that do not guarantee the optimal solution of the problem). It is this way that could lead to a significant effect of applying mathematical methods and computers to solve complex applied problems of discrete optimization.

The development and active use of various modifications of the method of sequential analysis of variants [17] in analyzing important economic problems gave rise to other general and special algorithms of solving multiple-choice problems.

In the early 1960s, in parallel with the method of sequential analysis of variants, I. V. Sergienko proposed [164] a new approach to develop algorithms for the approximate solution of discrete optimization problems of general form:

$$\min_{x \in G} f(x), \tag{2.29}$$

where G is a finite or countable set of feasible solutions of the problem and $f(x)$ is the objective function defined on this set. A new algorithm, so-called descent vector algorithm, was created. It makes it possible to develop a wide range of new local search algorithms and to select the most suitable one after the analysis of a specific problem.

The descent vector method may be classed among probabilistic local search methods due to the random choice of the initial approximation and enumeration of points in the neighborhoods.

Let us detail the descent vector method since it was well developed by Sergienko, his disciples, and various experts (e.g., [26, 124, 153, 156, 158, 159]) and was used to create new methods and various computer-aided technologies to solve many classes of discrete programming problems.

A point $x \in G$ will be called the point of local minimum of a function $f(x)$ with respect to a neighborhood $N(x)$ if this neighborhood is not degenerate (is not empty) and any of its points satisfies the condition $f(x) \leq f(y)$.

Let for $k = 1, ..., r$, $r \geq 1$, systems of neighborhoods $N^k(x)$ be given, such that $\forall x \in G \, N^j(x) \subseteq N^i(x)$ if $i, j \in \{1, ..., r\}$, $j \leq i$.

A vector function $\Delta^k(x)$ specified for each point $x \in G$ and neighborhood $N^k(x)$, $k = 1, ..., r$, is called the descent vector of the function f with respect to the neighborhood $N^k(x)$ if the following holds:

1. At each point $x \in G$, the value of the function $\Delta^k(x)$ is a q-measurable ($q = q(k,x) = |N^k(x)|$) vector with the coordinates $\Delta_1^k(x), \ldots, \Delta_q^k(x)$, which are real numbers, where $|N^k(x)|$ is the cardinality of the set $N^k(x)$.
2. $x \in G$ is a point of local minimum of the function f with respect to the neighborhood $N^k(x)$ if and only if $\Delta_i^k(x) \geq 0$ for all $i = 1, ..., q$.

3. If $x \in G$ is not a point of local minimum of the function f with respect to the neighborhood $N^k(x)$, then the descent vector can be used to find a point $y \in N^k(x)$ such that $f(y) < f(x)$.

By definition, vector $\Delta^k(x)$ allows finding the direction of decrease ("descent") of values of the function f in the neighborhood $N^k(x)$ for each point $x \in G$. The idea of using the descent vector $\Delta^k(x)$ to find such a direction underlies the method being considered.

In most cases, it is much easier to calculate the coordinates $\Delta_1^k(x), ..., \Delta_n^k(x)$ of the descent vector than the values of the function f at points of the neighborhood $N^k(x)$. Moreover, to find a point y in $N^k(x)$ such that $f(y) < f(x)$ or to establish that $f(x)$ is locally minimal with respect to this neighborhood, it will suffice to restrict the consideration to only some coordinates of the vector $\Delta^k(x)$.

The general scheme of the descent vector method can be presented as follows:

procedure **descent vector method**
  1. **initial_setting** $(s, x^s, s_{max}, \{I_k\}), 1 \leq k \leq r$
  2. *while* $(s \leq s_{max})$ do
  3. $k = 1$
      calculate_descent_vector $(\Delta^{l_k}(x^s))$
  4. **while** $(k \leq r)$ do
  5. **find_correction** $(\Delta^{l_k}(x^s), found)$
  6. if *found* = "FALSE" then
      comment the local minimum of the function f with respect to the neighborhood $N^k(x^s)$ has been found
  7. $k = k + 1$
  8. **calculate_descent_vector** $(\Delta^{l_k}(x^s))$
  9. else
10. $x^{s+1} = $ **new_point** $(N^k(x^s))$
11. $s = s + 1$
12. end if
13. end while
14. if $(k > r)$ break
15. end while
16. end

The procedure **initial_setting** specifies the initial data: the number of iterations $s = 0$, point $x^s \in G$, maximum number of iterations s_{max}, and a sequence $\{I_k\} = = I_1, ..., I_r$ of neighborhood numbers such that $I_1 < ... < I_r$.

The procedure **find_correction** searches for a negative coordinate of the descent vector $\Delta^{l_k}(x^s)$, $k \in \{1, ..., r\}$, $s \in \{0, ..., s_{max}\}$. A solution $y \in N^k(x^s)$ such that $f(y) < f(x^s)$ corresponds to this coordinate. After the procedure is executed, the variable *found* is assigned "TRUE" if a negative coordinate is found and "FALSE" otherwise.

The procedure **calculate_descent_vector** calculates (recalculates) the descent vector, and the procedure **new_point** uses some rule to choose a new solution $x^{s+1} \in N^k(x^s)$ such that $f(x^{s+1}) < f(x^s)$.

The algorithm of descent vector converges, which follows from the theorem proved in [153].

Only few steps of the algorithm require considering the neighborhood with the maximum number I_r. For most steps, it is sufficient to calculate the coordinates of the descent vector of the function $f(x)$ in the neighborhood $N^{I_1}(x^s)$. If neighborhoods with numbers larger than I_1 should be considered at some step, it will suffice to consider the set $N^{I_j}(x^s) \backslash N^{I_{j-1}}(x^s)$ instead of any such neighborhood $N^{I_j}(x^s)$, $j \in \{2, ..., r\}$, and to analyze only the coordinates of the descent vector $\Delta^{I_j}(x^s)$ that correspond to points of this set.

The algorithm of the descent vector is terminated once the number of steps has exceeded a preset value. It is also possible to assess how the value $f(x^s)$ obtained at each sth step meets practical requirements and to terminate the computational process if they are satisfied. The algorithm can also be terminated after some preset period t if the corresponding extremum problem is solved in a real-time decision-making system. In this case, naturally, the accuracy of the solution is strongly dependent on t.

In decision-support systems with participation of decision makers, they can stop the computational process based on informal considerations.

To complete the description of the descent vector method, we will dwell on the possible case where the local solution x^* found by the algorithm does not satisfy the researcher (user). To find other approximate solutions, the following ways are proposed:

1. Increase the neighborhood number I_r and continue the computational process according to the algorithm, based on the obtained point x^* as a new initial approximation. This procedure can be repeated several times. The feasible solutions produced by such procedures will obviously be better than x^*. Nevertheless, the computational process can be significantly complicated for large neighborhoods.
2. Increase the neighborhood number I_r and perform only one step of the algorithm to obtain point x' such that $f(x') < f(x^*)$. It is then recommended to come back to the initial neighborhood number I_1 and continue the computational process according to the algorithm.
3. Choose another initial approximation $x^{01} \in G$ and repeat computations based on this new approximation but with the same neighborhood number. The algorithm will generally result in another local solution. Repeated search for local minimum with a fixed neighborhood number and different initial approximations will obviously yield several local solutions; the solution is one that minimizes the function $f(x)$.
4. Use the following way: If there is a point x' in the neighborhood $N^{I_r}(x^*)$ such that $f(x') \neq f(x^*)$, then the algorithm can be continued by passing to the next step based on the point x'. Such a supplement to the basic algorithm does not usually involve intensive additional computations and, at the same time, can result in a more accurate solution in some cases.

Noteworthy is that the descent vector method was successfully used to solve many practical problems [124, 153, 156, 158, 159] formulated in terms of integer and combinatorial optimization. It is easy to use and combine it with other approximate algorithms.

The idea of regular change of the neighborhood of local search in the descent vector scheme was used in [216] to develop an efficient local-type algorithm with variable neighborhood. This algorithm and its modifications were successfully used to solve the traveling salesman and median problems and to find extremum graphs.

The GRASP algorithm [199], which fits well into the general scheme of the descent vector method, was first used to solve a complex graph covering problem and then other problems as well.

A school known to the scientific community was established at the V. M. Glushkov Institute of Cybernetics; it covers a wide range of modern problems in discrete optimization and systems analysis. Scientists of the Institute obtained fundamental scientific results in this field and analyzed applied problems for which real economic, engineering, social, and other characteristics and constraints are known. To solve such real problems, mathematical models and methods were developed; they allow obtaining solutions with a practicable accuracy and spending allowable amount of resources such as time and number of iterations.

These methods were incorporated into application software packages (ASP) created (and mentioned above in Chap. 1) at the Institute of Cybernetics and widely used to solve applied problems. They include DISPRO ASP for solving various discrete optimization problems [153], the PLANER ASP intended for high-dimensional manufacturing and transportation scheduling problems [116], etc. The experience gained in applying these packages underlies modern computer technologies.

As the application range widens, discrete programming models become more and more complicated [159, 160]. The dimension of actual applied discrete optimization problems that arise at the current stage exceeds by an order of magnitude the dimension of such problems solved earlier. This means that even the best discrete optimization methods fail to solve them, and the need arises to create essentially new mathematical methods that would be a combination, a system of interacting algorithms that purposefully collect and use information on the problem being solved and adjust the algorithm to the specific problem.

The recent studies carried out at the Institute of Cybernetics to develop new advanced probabilistic discrete optimization methods and compare them with the available ones based on a large-scale computational experiment on solving various classes of discrete optimization problems show the domination of the methods that combine various ideas developing within the framework of the local optimization theory. They helped to resolve a great many real-world challenges (pattern recognition, object classification and arrangement, design, scheduling, real-time control of processes, etc.).

One of such methods is the probabilistic global equilibrium search (GES) intended to solve complex discrete optimization problems [181]. It develops the descent vector method and uses the ideas of simulated annealing method. The GES method is based on repeated application of a "temperature" loop, where a series of iterations is carried out to choose points from the set of elite solutions and improve the obtained solutions using the tabu search algorithm. After finding a local optimum, the new method obtains a new initial point in a random way, the probabilities of its generation being determined by formulas similar to those used in the simulated annealing method.

Despite the simulated annealing method is successfully used to solve many complex optimization problems, its asymptotic efficiency is even lower than that of the trivial repeated random local search. The global equilibrium search has all the advantages, but not disadvantages, of the simulated annealing method. The results of numerous computational experiments on solving various classes of optimization discrete programming problems by the GES method [159, 160] suggest that this trend in the development of discrete optimization methods is promising.

To describe the global equilibrium search method, we will use the following formulation of a discrete optimization problem:

$$\min \{f(x) \mid x \in G \cap B^n\}, \tag{2.30}$$

where $G \subset R^n$, R^n is the set of n-measurable real vectors, and B^n is the set of n-measurable vectors whose coordinates take the values 0 or 1.

Denote $S = G \cap B^n$. Suppose $Z(\mu) = \sum_{x \in S} exp(-\mu f(x))$ for any $\mu \geq 0$. Define a random vector $\xi(\mu, \varpi)$:

$$\pi(x, \mu) = P\{\xi(\mu, \omega) = x\} = exp(-\mu f(x))/Z(\mu), \quad x \in S. \tag{2.31}$$

Distribution (2.31) is the Boltzmann distribution well known in statistical physics. It is related to simulated annealing method: Expression (2.31) determines the stationary probability of the fact that a Markov chain generated by this method is at the point x.

Let $S_j^1 = \{x \mid x \in S, x_j = 1\}$ and $S_j^0 = \{x \mid x \in S, x_j = 0\}$ be subsets of the set of feasible solutions of problem (2.30), where the jth, $j = 1, ..., n$, coordinate is equal to 1 and 0, respectively.

For $j = 1, ..., n$, determine

$$Z_j^1(\mu) = \sum_{x \in S_j^1} exp(-\mu f(x)), \quad Z_j^0(\mu) = \sum_{x \in S_j^0} exp(-\mu f(x)),$$

$$p_j(\mu) = P\{\xi_j(\mu, \omega) = 1\} = \frac{Z_j^1(\mu)}{Z(\mu)}.$$

Suppose that the set $\tilde{S}$ is a subset of the set of feasible solutions of problem (2.30) found by the GES method and

$$\tilde{S}_j^1 = \{x|\ x \in \tilde{S}, x_j = 1\},\ \tilde{S}_j^0 = \{x|\ x \in \tilde{S}, x_j = 0\},\ j = 1, ..., \ n.$$

Let us present the general GES scheme.

procedure GES
1. **initialize_algorithm_parameters** $(maxiter, ngen, maxnfail, \{\ \mu_k\}, 0 \le k \le K)$
2. while (stoppage_criterion $=$ FALSE) do
3. if $(\tilde{S} = \emptyset)$ then $\{\tilde{S}$ is the set of known solutions$\}$
4. $x \leftarrow$ **construct_solution**
   comment constructing a random feasible solution x
5. $g(x) \leftarrow$ **construct_decay_vector**(x)
6. $x_{max} = x;\ x_{best} = x;\ g_{max} = g(x)$
7. $\tilde{S} \leftarrow x_{max}$
8. $Elite \leftarrow x_{max}$ $\{Elite$ is the set of elite solutions$\}$
9. end if
10. $nfail \leftarrow 0$ $\{nfail$ is the number of full loops without correction$\}$
11. while $(nfail < maxnfail)$ do
12. $k \leftarrow 0$
13. $old\_x_{max} \leftarrow x_{max}$
14. while $(k < K)$ do
15. **calculate_probabilities_of_generation** $(pr^k, \tilde{S})$
16. $t \leftarrow 0$
17. while $(t < ngen)$ do
18. $x \leftarrow$ **generate_solution** (x_{max}, pr^k)
19. $R \leftarrow$ **search_method** (x)
   comment R is the set of found feasible solutions
20. $R \leftarrow R \backslash P$ $\{P$ is the set of infeasible solutions$\}$
21. $\tilde{S} \leftarrow \tilde{S} \cup R$
22. $x_{max} \leftarrow argmin f(x)$
23. $g(x_{max}) \leftarrow^{x \in \tilde{S}}$ **calculate_descent_vector** (x_{max})
24. if $(f(x_{max}) < f(x_{best}))$ then
25. $x_{best} \leftarrow x_{max}$
26. end if
27. **form_elite** $(Elite,\ R)$
28. $t \leftarrow t + 1$
29. end while
30. $k \leftarrow k + 1$
31. end while
32. if $(f(old\_x_{max}) = f(x_{max}))$ then $nfail \leftarrow nfail + 1$
33. else $nfail \leftarrow 0$
34. $\tilde{S} \leftarrow Elite$
35. end while
36. $P = P \cup N(x_{best},\ d_p)$
37. $Elite = Elite \backslash P$
38. if (RESTART-criterion $=$ TRUE) then $Elite \leftarrow \oslash$

39. $\tilde{S} \leftarrow Elite$
40. $x_{max} = argmin f(x)$
41. $g(x_{max}) \leftarrow^{x \in \tilde{S}}$ **calculate_descent_vector** (x_{max})
42. end while
end GES

In developing GES-based algorithms, use is made of rules that allow:

- Using the information obtained in the problem solution
- Intensifying and diversifying the search of an optimum
- Using the principles of RESTART technology [159]
- Combining different metaheuristics and forming a "team" of algorithms for the optimal use of the advantages of the algorithms and the specific features of the problem

Let us detail the design of GES algorithms based on these rules. The outer loop (iteration loop, lines 2–42) is intended for repeated search for the optimal solution. Together with the operators in lines 37–39, it allows using the RESTART technology. This loop either performs a prescribed number of repetitions or is executed until some stopping criterion. The stopping criterion may be, for example, a solution with a value of the objective function better than a known record. An essential element of the GES algorithm is the "temperature" loop (lines 14–31), which needs the values of K and "temperature" μ_k, $k = 0, \ldots, K$, to be specified. This loop starts a series of searches for the optimal solution for increasing "temperatures." The "temperature" loop and its repetitions allow flexible alternation of narrowing and expanding of the search zone, which eventually contributes to the high efficiency of the GES method.

If the set $\tilde{S}$ of solutions known at some moment is empty, the operators of lines 4–9 allow finding the first solution and initializing the necessary data. Note that the solutions found by the GES algorithm are not saved (except for elite solutions, which appear in the set *Elite*) and are used to calculate the quantities

$$\tilde{Z}_k = \sum_{x \in \tilde{S}} exp\{-\mu_k f(x)\}, \quad \tilde{F}_k = \sum_{x \in \tilde{S}} f(x) exp\{-\mu_k f(x)\}, \quad \tilde{E}_k = \frac{\tilde{F}_k}{\tilde{Z}_k},$$

$$\tilde{Z}_{kj}^l = \sum_{x \in \tilde{S}_j^l} exp\{-\mu_k f(x)\}, \quad \tilde{F}_{kj}^l = \sum_{x \in \tilde{S}_j^l} f(x) exp\{-\mu_k f(x)\}, \quad \tilde{E}_{kj}^l = \frac{\tilde{F}_{kj}^l}{\tilde{Z}_{kj}^l},$$

$$k = 0, \ldots, K, \quad l = 0, 1, \quad j = 1, \ldots, n. \tag{2.32}$$

They allow determining the probabilities $\tilde{p}^k = (\tilde{p}_1^k, \ldots, \tilde{p}_n^k)$ according to one of the following formulas (see, e.g., [159]):

$$\tilde{p}_j^k = \tilde{p}_0^k exp\left\{\frac{1}{2} \sum_{i=0}^{k-1} (\tilde{E}_i^0 + \tilde{E}_{i+1}^0 - \tilde{E}_{ij}^1 - \tilde{E}_{i+1j}^1)(\mu_{i+1} - \mu_i)\right\}, \tag{2.33}$$

$$\tilde{p}_j^k = \cfrac{1}{1 + \cfrac{1-\tilde{p}_0^k}{\tilde{p}_0^k} exp \left\{ -0.5 \sum_{i=0}^{k-1} (\tilde{E}_{ij}^0 + \tilde{E}_{i+1j}^0 - \tilde{E}_{ij}^1 - \tilde{E}_{i+1j}^1)(\mu_{i+1} - \mu_i) \right\}}, \qquad (2.34)$$

$$\tilde{p}_j^k = \tilde{Z}_{kj}^1 / \tilde{Z}_k. \qquad (2.35)$$

The probabilities $\tilde{p}^k$ of generating the initial solutions (line 18) for the **search_method**(x) procedure (line 19) are estimated based on the concepts taken from the simulated annealing method and depend on the current "temperature" μ_k and the set $\tilde{S}$ of feasible solutions found earlier.

The values of μ_k, $k = 0, ..., K$, determine the annealing curve and are usually estimated by the formulas $\mu_0 = 0$, $\mu_{k+1} = \sigma\mu_k$, $k = 1, ..., K - 1$. The values of μ_1 and $\sigma > 1$ are chosen so that the probability $\tilde{p}^K$ is approximately equal to $x_{max} \| \| x_{max} - \tilde{p}^K \| \cong 0$. Note that the annealing curve is universal and is used to solve all problems rather than a specific one. Only the coefficients of the objective function are scaled so that the record is equal to a prescribed value. The initial probability $\tilde{p}^0$ can be specified in different ways: All the coordinates of the vector $\tilde{p}^0$ are identical (e.g., 0.5), this vector is a solution to problem (2.30) with the relaxation of Boolean conditions or it is obtained by statistical modeling of a Markov chain, which corresponds to the simulated annealing method for problem (2.30) with infinite temperature (all the feasible solutions are equiprobable).

The correcting loop (lines 11–34) is carried out until the solution x_{max} is improved *maxnfail* times. The procedure **calculate_probabilities_of_generation** (line 15) is used to estimate the probabilities of random perturbation of the solution by one of the formulas (2.33)–(2.35). The loop of finding new solutions (lines 17–29) repeats a prescribed number of times *ngen*.

Let us present the **generate_solution** procedure.

procedure **generate_solution** (x_{max}, pr^k)
1. $x \leftarrow x_{max}$; $dist = 0$; $j \leftarrow 1$
2. while $(j \leq n)$
3. if $x_j = 1$ then
4. if $(\tilde{p}_j^k < random[0, 1])$ then
5. $x_j \leftarrow 0$; $dist \leftarrow dist + 1$
6. $g(x) \leftarrow$ **recalculate_descent_vector** (x)
7. end if
8. end if
9. else
10. if $(\tilde{p}_j^k \geq random[0, 1])$ then
11. $x_j \leftarrow 1$; $dist \leftarrow dist + 1$
12. $g(x) \leftarrow$ **recalculate_descent_vector** (x)
13. end if
14. end else
15. $j \leftarrow j + 1$
16. if $(dist = max\_dist)$ then break
17. end while

This procedure randomly generates the solution x, which is initial for the **search_method** procedure (line 19). For low "temperatures" μ_k, the perturbation of the solution x_{max} is purely random and equiprobable; for high ones, the values of the coordinates that are identical in the best solutions vary slightly. The "temperature" loop allows diversifying the search (for low "temperatures") and intensifying it (for high "temperatures"). Thus, the solutions generated as the "temperature" increases get features inherent in the best solutions, and as a result, they coincide with the solution x_{max}.

In line 1, the initial values are assigned to the vector x and variables $dist$ and j. The vector x_{max} is perturbed in the loop in lines 2–17. The quantity $random[0, 1]$ uniformly distributed on the interval $[0, 1]$ (lines 4 and 10) is used to model random events that change the solution x_{max}. Noteworthy is that the descent vector $g(x)$ is recalculated after each perturbation of x_{max}. It is specified by an n-measurable vector defined on the neighborhood of unit radius with the center at the point x: $g_j(x) = f(x_1, \ldots, x_{j-1}, 1 - x_j, x_{j+1}, \ldots, x_n) - f(x_1, \ldots, x_j, \ldots, x_n)$.

Recalculating the vector $g(x)$ corresponds to the idea of the descent vector method and saves much computational resources for many problems. For example, for an unconstrained quadratic-programming problem with Boolean variables, calculating the objective function requires $O(n^2)$ arithmetic operations, and recalculating the vector $g(x)$ needs only $O(n)$ such operations.

The perturbation loop (lines 2–17) is terminated if the Hamming distance between the point x_{max} and the perturbed solution x is equal to a prescribed parameter max_dist (line 16). This parameter can be specified dynamically and depend on a "temperature" index k. Usually, max_dist is chosen on the interval $\left[\frac{n}{10}, \frac{n}{2}\right]$.

In solving different problems, the **search_method** procedure employs randomized (probabilistic) algorithms of local search, descent vector, tabu search, and simulated annealing. When an algorithm is chosen for the procedure, specific features of the problem being solved are taken into account. The algorithm should intensify the search and allow the efficient study of the neighborhoods of the initial solution in order to correct it. Moreover, the algorithm can be chosen dynamically, that is, one algorithm can be used for low "temperatures" and another for high ones.

The **search_method** procedure is used to find the set of local (in the sense of a neighborhood of certain radius) solutions that form the set R. Lines 20–21 formally specify that the set R has no currently infeasible solutions and that quantities (2.32) are recalculated. In the GES algorithm, the search for infeasible solutions is blocked, and recalculation is carried out as new solutions are found.

The solution x_{max} and the descent vector at this point are found in lines 22 and 23, respectively. The **form_elite** procedure specifies the set $Elite$ of elite solutions, which consists of solutions with the best values of the objective function: $minf(x)_{x \in Elite} \geq max\, f(x)_{x \in \bar{S} \backslash Elite}$. The maximum cardinality of the set is determined by the parameter max_elite_size and is usually chosen from the interval $[1, n]$.

After the correction loop (lines 11–35) is terminated, the set P of infeasible points is supplemented with points from a neighborhood centered at x_{max} specified by the parameter d_P (line 36), and infeasible points are removed from the set $Elite$ (line 37).

If the analysis of the optimization process detects the so-called RESTART distribution of the time of finding solutions, the set *Elite* is made empty, which allows restarting a new independent solution process.

The GES method was extended to the following classes of discrete optimization problems: a multidimensional knapsack problem with Boolean variables, problems on the maximum cut of an oriented graph, maximum feasibility, scheduling, finding a clique, graph coloring, quadratic programming with Boolean variables, etc. The efficiency of the developed GES algorithms was compared with that of known algorithms [159, 160]. An analysis of the algorithms and the results they produced shows that for all the problems solved, the global equilibrium search has significant advantages over the other methods, which demonstrate high efficiency for one problems and very low efficiency for others. The GES algorithm performed the best for all the subsets of problems, including the above classes of complex high-dimensional discrete programming problems, and successfully competed with the approximate algorithms that are currently the best.

Noteworthy is that the main advantage of the GES method is not its computational efficiency but rather its team properties (the ability to make efficient use of the solutions obtained by other algorithms). For example, the classical local search method has zero team index. The team properties allow the GES method to be efficiently used for parallel computing, which is especially important for the optimization in computer networks and the Internet. In our opinion, in developing optimization methods of new generation in the future, optimization should be interpreted not in the commonly accepted sense but as a team of algorithms that work in parallel, acquiring and exchanging information.

Recent developments also include a theoretical basis for accelerated solution of complex discrete optimization problems. In particular, the above-mentioned RESTART technology [159] was developed; it is based on the new concepts of RESTART distribution and RESTART stopping criterion. RESTART distribution is the distribution of time necessary to solve the problem such that the average time can be reduced by using a restarting procedure. The idea of the RESTART technology is to randomize and modify an optimization algorithm so as to make the distribution of the time it solves the problem a RESTART distribution for which the use of the optimal RESTART stopping criterion considerably accelerates the problem solution.

Procedures for choosing optimization algorithms based on the method of probabilistic decomposition and the Wald criteria of sequential analysis were developed and analyzed.

Supercomputers created in the recent years due to the rapid development of computer and communication facilities determined one of the possible ways to accelerate the solution of complex and labor-intensive discrete programming problems and required new algorithms for parallelizing the optimization process.

The new approach to parallelizing the optimization process proposed in [159] is promising. Instead of the operations executed by the algorithm, it parallelizes its interacting copies.

Theoretical studies and operation paralleling were used to develop an information technology to solve problems of constructing maximum-size barred codes.

Such problems have attracted interest because of the development of information technologies, the Internet, computer networks, and modern communication facilities. The developed information technology allowed obtaining new record results: Using the SCIT-3 multiprocessor complex, maximum-size barred codes were constructed for the first time or the records existing for them were improved [159, 160]. To get acquainted with them, see http://www.research.att.com/~njas/doc/graphs.html (research center of computer science AT&T Labs, New Jersey, USA), where the global level of achievements in this field is shown.

Problems of correctness and stability of multicriterion (vector) problems [90, 157, 159] occupy an important place in the studies carried out in the last decades by experts in discrete optimization, including scientists at the Institute of Cybernetics. The concept of a correct mathematical problem is closely related to the following properties: The problem should be solvable and have a unique solution, and this solution should continuously depend on the variations in input data. The third property is usually called stability. Close attention to the correctness of optimization problems is largely due to the fact that in solving many applied problems that can be formalized by mathematical optimization models, it is necessary to take into account uncertainty and random factors such as the inaccuracy of input data, mathematical models inadequate to real-world processes, roundoff errors, errors of numerical methods, etc. Studies carried out by scientists at the Institute were mainly intended to establish the conditions whereby the set of optimal solutions (Pareto, Slater, or Smale set) of a vector discrete optimization problem has any of the five properties, each characterizing in a certain way the stability of the problem against small perturbations of input data. Omitting the necessary definitions and formalized description of the results obtained, we will specify only the main ones. Five types of problem stability were considered. The regularization of unstable problems was investigated. The stability of a vector integer optimization problem on a finite set of feasible solutions was related to the stability of optimal and non-optimal solutions. The concepts that may underlie the description of different types of stability to formulate necessary and sufficient stability conditions were defined. It was proved that different types of stability of the problem can be ascertained on a common basis by studying two subsets of a feasible set, namely, all the points that stably belong to the set of optimal solutions and all the points that do not stably belong to it. A common approach to the analysis of different types of stability of a vector integer optimization problem was proposed. The obtained results are related to the stability of both the perturbations of all input data of the problem and the perturbations of the input data necessary to represent its vector criterion or constraints, which is important since a problem stable against perturbations of some input data may be unstable against perturbations of the remaining data.

The theoretical and applied studies in discrete optimization carried out at the Institute of Cybernetics provide a basis for various information technologies.

2.6   Results in Applied Systems Analysis and Professional Training in this Field

V. S. Mikhalevich paid special attention to methods of systems analysis. In the last 15–20 years, they have been substantially developed by scientists from many organizations, including the Institute of Applied Systems Analysis of the National Academy of Sciences of Ukraine and Ministry of Education and Science of Ukraine (IASA). The ideological fundamentals of interdisciplinary studies and scientific developments combined with the educational process turned out to be important for both scientific research and training of students who have come to be often engaged in real research. More complex interdisciplinary problems to be solved necessitate the creation of a system methodology and the development of the theory and practice of systems analysis. This necessity is due to not only the intensive development of science and technology but also constant accumulation of threats of ecological, technogenic, natural, and other disasters.

Fundamental and applied research intended to develop a methodology of systems analysis of complex interrelated social, economic, ecological, and technological objects and processes go to the front burner more than ever today. A methodology to solve this class of problems is a new field of studies called system mathematics. It is a complex of interrelated subdisciplines (classical and new ones) of mathematics that allow solving various modern interdisciplinary problems. For example, elements of system mathematics are a formalization of the interrelations between continuous and discrete mathematics, transformation of some optimization methods into fuzzy mathematics and development of appropriate optimization methods in fuzzy formulation, description of interrelated processes that develop in different time scales (with different rates), analysis of distributed- and lumped-parameter systems on a unified platform, and combination of methods of quantitative and qualitative analysis in unified computational processes in constructing man–machine systems, etc.

The department of mathematical methods of systems analysis (MMSA), which is a part of the IASA, was created, as already mentioned, at the NTUU "KPI" in 1988 on Mikhalevich's initiative. Five years later, Academician Kukhtenko and Mikhalevich participated in creating the Research Institute of Interdisciplinary Studies (RIIS) on the basis of the research sector of MMSA department at the Kyiv Polytechnic Institute. It was this department and the RIIS and two research departments of the V. M. Glushkov Institute of Cybernetics supervised, respectively, by B. N. Pshenichnyi, academician of the National Academy of Sciences of Ukraine, and V. S. Mel'nik, corresponding member of the National Academy of Sciences of Ukraine, that became the basis for the Scientific and Training Complex "Institute for Applied System Analysis."

M. Z. Zgurovsky, academician of NAS of Ukraine

In what follows, we will dwell on some important developments in Applied Systems Analysis carried out at the IASA under the guidance of M. Z. Zgurovsky. These scientific studies prove that it is expedient to use system methodology to solve modern interdisciplinary problems and substantiate the role and place of systems analysis as a universal scientific methodology in science and practice and its interrelation with other fundamental disciplines. The key concepts, axioms, and definitions of systems analysis were formulated. The concept of complexity as a fundamental property of problems of systems analysis and principles and methods of solving problems of transcomputational complexity were formulated within the framework of the general problem of systems analysis [70, 72, 225]. A constructive and convenient way of representing the initial information on the object as conceptual spaces of conditions and properties of the object was proposed to solve complex interdisciplinary problems. Under modern conditions, these spaces should provide new vision of element interaction in the structure "system analyst ⇔ the human ⇔ object ⇔ environment." This will provide more clear presentation of the coordination of major factors: properties of the object under study, input information, and operating conditions in view of various uncertainties and multifactorial risks. The coordination should be systemic and take into account the objectives, problems, expected results of operation of the object, the complexity of situations in which it operates, and the shortage of information on the complexities related to the objectives and operation conditions.

In order to create tools to solve complex interdisciplinary problems, approaches are developed, which help to formalize the procedures of conceptual–functional spaces of conditions and properties of system operation. Under these conditions, problems of evaluating indeterminate forms of objectives, situations, and conflicts in problems of interaction and counteraction of coalitions are formulated and methods are proposed to solve them. Methods of evaluating indeterminate forms of objectives are developed: using technical constraints, reducing to a system of nonlinear incompatible equations, and reducing to a Chebyshev approximation problem. The problems of evaluating natural and situational indeterminate forms were considered: evaluating indeterminate forms for known characteristics of random factors, evaluating indeterminate forms in the case of incomplete information on random factors. Problems of evaluating indeterminate forms in conflict situations were analyzed: evaluation of indeterminate forms of active interaction of partners and counteraction of opposite parties, multi-objective interaction of partners under situational uncertainty, and multi-objective active counteraction of opposite parties under situational uncertainty.

The approaches to searching for a rational compromise in evaluating conceptual indeterminate forms and an approach to evaluating functional dependences in these problems were proposed. The problem of evaluating functional dependences is essentially more complex than the typical problem, which is due not only to heterogeneous input information but also heterogeneous properties of the groups of factors considered. To overcome the transcomputational complexity, it is expedient to generate approximation functions as a hierarchical multilevel system of models. Within the conceptual–functional space of properties, it is expedient to analyze the issues related to constructing the structure and functions of complex multilevel hierarchical systems and to formulate the principles and methods of structuring the formalized description of properties, structures, and functions of such class of systems. To give the mathematical formulation of the problem of systems analysis of a complex multilevel hierarchical system, the general solution strategy and structure of the generalized algorithm of the structural–functional analysis, as well as the method of its solution are proposed [70, 72, 225]. An approach to solving the problem of system structural optimization of complex structural elements of modern technology is proposed, which is based on a purposeful choice of functional elements of each hierarchical level. Problems of system parameter optimization are solved, which provides a rational compromise of inconsistent requirements to the strength, reliability, manufacturability, and technical and economic efficiency of the structure.

Professor N. D. Pankratova

Based on methods of system mathematics and probability theory, functional analysis, function approximation theory, and discrete mathematics, Pankratova proposed [133] a new principle of before-the-fact prevention of the causes of possible transition of an object from the operable to a faulty state, based on systems analysis of multifactor risks of contingency situations, reliable estimation of the resources of admissible risk of different operation modes of a complex engineering object, and prediction of the key "survivability" parameters of the object during its operation life. An apparatus is proposed for a system-coordinated solution of problems of detecting, recognizing, predicting, and minimizing risks of contingency, critical, and emergency situations, accidents, and catastrophes.

Let us formulate the problem of detecting a contingency situation in the dynamics of operation of an ecologically dangerous object [133].

Given: for each situation $S_k^\tau \in S_\tau$, a set $M_k^\tau \in M_\tau$ of risk factors is formed as $M_k^\tau = \{\rho_{q_k}^\tau | q_k = \overline{1, n_k^\tau}\}$. For each risk factor of the set M_k^τ, fuzzy information vector $I_q^\tau = \{I_{q_k}^\tau | q_k = \overline{1, n_k^\tau}; k = \overline{1, K_\tau}\}$ is known and its components are known to have the form

$$I_{qk}^\tau = \{\tilde{x}_{q_k j_k p_k}^\tau | q_k = \overline{1, n_k^\tau}; j_k = \overline{1, n_{q_k}^\tau}; p_k = \overline{1, n_{q_k j_k}^\tau}\},$$

$$\tilde{x}_{q_k j_k p_k}^\tau = \langle x_{q_k j_k p_k}^\tau, \mu_{H_{q_k j_k p_k}}(x_{q_k j_k p_k}^\tau) \rangle, \quad x_{q_k j_k p_k}^\tau \in H_{q_k j_k p_k}^\tau, \quad \mu_{H_{q_k j_k p_k}} \in [0, 1],$$

$$H_{q_k j_k p_k}^\tau = \langle x_{q_k j_k p_k}^\tau | x_{q_k j_k p_k}^- \le x_{q_k j_k p_k}^\tau \le x_{q_k j_k p_k}^+ \rangle.$$

Required: for each situation $S_k^\tau \in S_\tau$ and each risk factor $M_k^\tau \in M_\tau$, recognize a contingency situation in the dynamics of operation of an ecologically dangerous object and ensure the "survivability" of a complex engineering system during its operation.

The problem of diagnostics of the operation of a deep pumping facility is used to show that it is expedient to apply methods that allow before-the-fact prevention of contingency situations, that is, ensure the "survivability" of the complex system. The main purpose of the pumping facility is to maintain a prescribed level of water delivery for various needs, cooling a technological ecologically dangerous installation being of priority.

A systemic approach is employed to solve problems that are of current importance for the economy and society. A system methodology of developing scenarios of future events in various spheres of human activity with the use of unique man–machine tools of technological forecast as an information platform of scenario analysis is developed [71]. This technology is a complex of mathematical, program, logic, and technical–organizational means and tools to provide an integral process of forecasting based on the interaction of man and specially created software-engineering environment that allows monitoring the decision-making process and choosing a rational compromise based on proposed alternative scenarios.

Based on the developed approach to the solution of problems of technological forecast, a number of projects were implemented for Ministry of Economics, Ministry of the European Integration of Ukraine, Ministry of Education and Science of Ukraine, National Space Agency of Ukraine, and Scientific Center for Aerospace Research of the Earth at the Institute of Geological Sciences of the National

Academy of Sciences of Ukraine. In particular, by the order of the National Space Agency of Ukraine, priority fields of the consumption of space information provided by Earth remote sensing systems were identified. For the integrated iron-and-steel plant in Krivoi Rog, alternative scenarios for increasing the efficiency of the logistic system of the plant as to maintaining all the stages of metallurgical production were created at the stage of short-term forecast. By the order of the State Agency of Ukraine on Investments and Innovations, a strategic plan for the development of the Autonomous Republic of Crimea and Sevastopol were developed.

The methodological and mathematical principles and approaches to implementing the strategy of technological forecast can be used to predict the technological development of the society. This approach can be applied to develop scenarios for the innovation development of large enterprises and individual branches of industry and to form the technological policy of the society.

Based on the methods of systems analysis, Zgurovsky obtained fundamental results in global modeling of processes of sustainable development in the context of the quality and safety of human life and the analysis of global threats [69]. A system of indices and indicators was generated and a new metrics was proposed to measure processes of sustainable development in three spaces: economic, ecological, and social (institutional). This metrics was used to perform the global modeling of processes of sustainable development for a large group of countries. An important division of modeling deals with the analysis of the pattern of world conflicts as a fundamental property of the global society. An attempt was made to predict the next world conflict called "the conflict of the twenty-first century," and its nature and key characteristics (duration, main phases, and intensity) were analyzed. The set of main global threats that generate this conflict was presented. The method of cluster analysis was used to determine how these threats influence different countries and eight large groups of countries (civilizations by Huntington) with common culture. The assumptions on possible scenarios of the "conflict of the twenty-first century" were suggested.

The reliability of any forecast "resonating" with other global or local tendencies, hypotheses, and patterns is known to substantially increase. For such additional conditions, Zgurovsky took the modern concept of the acceleration of historical time and the hypothesis that the duration of large K-cycles decreases with scientific and technical progress [67]. Starting from these and considering the evolutionary development of civilization as an integral process determined by the harmonic interaction of its components, the Kondratieff cycles of the development of the global economy were compared with the C-waves of global systemic conflicts, and an attempt was made to predict these periodic processes for the twenty-first century. Analyzing these two processes on a common time axis reveals their agreement (synchronism), which is formulated as the following two principles.

1. *Quantization Principle*. Time intervals $\Delta(C_n)$, $n \geq 5$, during which the wave C_n passes five evolutionary phases "origin $\to$ growth $\to$ culmination $\to$ decrease $\to$ decay" contain an integer number $n_k(\Delta(C_n))$ of full K-cycles of a modified sequence of Kondratieff cycles (KCs) $\{K_n\}_{n \geq 1}$.

2. *Monotonicity Principle.* The average duration $T_k(\Delta(C_n))$ of one full K-cycle of the KC $\{K_n\}_{n\geq 1}$ on the time intervals $\Delta(C_n)$ decreases as n grows.

Denote by

$$G(C_k; \{ K_n\}_{n\geq 1}) \stackrel{\Delta}{=} \{K_{s(k)}; K_{s(k)+1}; \ldots ; K_{s(k)+m(k)}\}, \quad k \geq 5,$$

a group (quantum) of K-cycles separated by the C_k-wave from the KC $\{K_n\}_{n\geq 1}$. Then,

$$n_k(\Delta(C_k)) = m(k) + 1; \quad T_k(\Delta(C_k)) = (m(k) + 1)^{-1} \times \sum_{r=0}^{m(k)} T(K_{s(k)+r}),$$

where $T(K_j)$ is the duration of one full K-cycle K_j.

In our case,

$$G(C_5; \{K_n\}_{n\geq 1}) = \{K_1; K_2; K_3\}, \quad G(C_6; \{ K_n\}_{n\geq 1}) = \{ K_4; K_5\},$$

$$T_k(\Delta(C_5)) = 3^{-1} \sum_{i=1}^{3} T(K_i) = 56.6 \text{ years}, \quad n_k(\Delta(C_5)) = 3,$$

$$T_k(\Delta(C_6)) = 2^{-1} \sum_{i=4}^{5} T(K_i) = 43.5 \text{ years}, \quad n_k(\Delta(C_6)) = 2.$$

The pattern revealed allows formulating the hypothesis on the next quantization step, that is, separating the next group $G(C_7; \{K_n\}_{n\geq 1})$ of K-cycles from the KC $\{K_n\}_{n\geq 1}$ by the C_7-wave.

Since the development of the global economy and the course of systemic global conflicts are interdependent components of the same process of evolution of the globalized society, the agreement (synchronism) of these processes revealed on the time intervals $\Delta(C_5)$ and $\Delta(C_6)$ in the sense of quantization and monotonicity principles remains on the time interval $\Delta(C_7)$.

Based on the basic hypothesis, the course (in the metric aspect) of K-cycles in the twenty-first century is forecasted as follows:

(a) The time interval $\Delta(C_7)$ contains no less than two full cycles of the KC $\{K_n\}_{n\geq 1}$.
(b) The average duration of one full K-cycle on the time interval $T_k(\Delta(C_7))$ is shorter than $T_k(\Delta(C_6)) = 43.5$ years.

Thus, two cases are possible, which correspond to two scenarios of the course of Kondratieff cycles in the twenty-first century.

1. The time interval 2008–2092 contains two full Kondratieff cycles:

$$G(C_7; \{K_n\}_{n\geq 1}) = \{K_6; K_7\}, \quad n_k(\Delta(C_7)) = 2,$$

$$T_k(\Delta(C_7)) = 2^{-1} \sum_{i=6}^{7} T(K_i) = 42.5 \text{ years} < T_k(\Delta(C_6)) = 43.5 \text{ years}.$$

2. The time interval 2008–2092 contains three full Kondratieff cycles:

$$G(C_7; \{K_n\}_{n\geq 1}) = \{K_6; K_7; K_8\}, \quad n_k(\Delta(C_7)) = 3,$$

$$T_k(\Delta(C_7)) = 3^{-1} \sum_{i=6}^{8} T(K_i) = 28.3 \text{ years} < T_k(\Delta(C_6)) = 43.5 \text{ years}.$$

The results of the study confirm the refined hypothesis that the duration of large Kondratieff cycles tends to decrease with the scientific and technical progress: It is most probable that in the twenty-first century, there will appear three large K-cycles with about 30-year average duration of one full cycle, which is much shorter than the average duration of one of the previous five Kondratieff cycles ($\approx$50 years). Moreover, the revealed synchronization of the development of the global economy and of the course of global systemic conflicts may be interpreted as an indirect confirmation that the models of Kondratieff cycles and C-waves are correct.

Methods of the optimal control of nonlinear infinite-dimensional systems developed at the IASA were used to develop a fundamentally new approach to the analysis of singular limiting problems for equations of mathematical physics. This approach is based on reducing an ill-posed problem to an auxiliary (associated) optimal control problem. Following this way, it is possible to develop stable solution algorithms for three-dimensional Navier–Stokes equations, Bénard systems, etc. Thus, methods of optimal control theory are a mathematical tool for efficient analysis of ill-conditioned equations of mathematical physics.

A number of results were obtained in nonlinear and multivalued analysis, solution of nonlinear differential-operator equations, analysis of inclusions and variational inequalities as an important field of the development of methods of infinite-dimensional analysis, theory and methods of optimization, game theory, etc. [91, 92, 205, 214]. A new unified approach was developed to solve optimal control problems using systems of nonlinear processes and fields. New results were obtained in the theory of control of complex nonlinear dynamic systems that contain modules with distributed and lumped parameters.

Studies on constructing and analyzing models of social systems were performed recently at the IIASA. One of the fields was constructing network-type models with associative memory (similar to Hopfield neural networks) for a wide class of processes: social, economic, political, and natural. We may say that revealing the multiple-valued behavior of the solutions of such models is an achievement in this field. This opens up a way to the wide application of methods of system mathematics. For example, there arise problems of the analysis of model behavior depending on parameter variations in the case of multivaluedness, limit cycles, and especially in the case of multivalued mappings (the groundwork for their analysis was laid earlier). Solving these problems requires almost the whole arsenal of means of system mathematics: nonlinear functional analysis, dynamic systems theory, and theory of differential and discrete equations.

One more field of studies is related to modeling the motion of a great number of objects. The cellular automata method was chosen as one of the approaches. It analyzes the motion within the framework of discrete approximation (the space being divided into regular cell unions) and considers the passages of objects from one cell to another. Research of cellular automata with prediction was launched (O. S. Makarenko). Mathematically, this problem looks like studying discrete dynamic systems (deterministic or stochastic). Traffics were modeled under some conditions with the use of models adapted to real geometry of transportation flows and obstacles, and problems on the allowance for motion prediction by the participants were also considered. Such problems again necessitate the mathematical analysis based on optimization and multivalued mapping theories.

Dynamic and time characteristics of some macroeconomic and social processes as fundamentals for their analysis and prediction were found out and investigated. The following problems were formulated and solved based on statistical information: qualitative analysis of the stages of development, growth, decrease, and decay of social processes; analysis of conceptual relationships among the dynamic and time characteristics of each individual stage of a process; formalization of the dynamic and time characteristics of processes; creation of computation algorithms based on statistical data; and testing the results of the analysis. Specific features of "living" systems are the conceptual basis of the analysis. The theoretical basis is the dynamic systems theory, nonequilibrium states theory, self-organization theory, and systems analysis (M. V. Andreev, G. P. Poveshchenko). Applied mathematical models for short- and intermediate-term forecast of the key macroeconomic indices of Ukraine (rate of gross domestic product, rate of employment of the population, rate of consolidated budget, rate of expenses, etc.), a mathematical model of labor market, and a mathematical model of the structural evolution of productive forces were created and tested.

The system methodology in optimization, in particular, system optimization as a purposeful modification of the feasible domain of complex engineering and economic optimization problems, is developed at the department of numerical optimization methods of the IASA. The subject of research is mathematical models that describe flows in networks with generalized conservation law, where the difference of inflows and outflows at a node may be not equal exactly to the volume of consumption at the node but belong to some interval. Such a model occurs, for example, in control problems of water transfer in irrigation channels. Flow calculation problems consider not only arc but also nodal variables. Flow transportation costs significantly decrease in this case due to optimal redistribution of sources (drains). Nonlinear complex models are constructed, which describe resource transportation and distribution among consumers in view of the possibility of creating inventory in certain tanks (storehouses) such as gas flow distribution models under active operation of gasholders [132].

In his scientific studies, Mikhalevich paid much attention to modeling and prediction. During the last two decades, scientists of the IASA fruitfully worked to develop methods of mathematical and statistical modeling and forecast of arbitrary processes. Methods of regression data analysis received further development and original structures of models were obtained to describe nonlinear nonstationary processes (e.g., models of heteroscedastic and cointegrable processes, which are popular in the finance and economy). An original technique was created to choose the type and to evaluate the structure of a model and is successfully used in creating decision-support systems. Computational procedures were proposed for data preprocessing in order to reduce them to a form that is most convenient to estimate the parameters of mathematical, including statistical, models. To overcome difficulties due to different types of uncertainties, probabilistic modeling and prediction methods were developed, which use the Bayesian approach to data analysis, in particular, Bayesian networks [68]. The advantages of models in the form of Bayesian networks are their clearness (directed graph), the possibilities of creating highly dimensional models (in the sense of a large number of nodal variables), the use of continuous and discrete variables in one model, allowance for uncertainties of structural and stochastic types, and possible use of methods of exact and approximate probabilistic inference. Bayesian networks can generally be characterized as a rather complex resource-intensive yet highly effective probabilistic method to model and predict the development of various processes.

Processes with uncertainties were predicted in the best way with the use of probabilistic methods and fuzzy logic. Inherently, these methods are close to situation modeling and decision making by man; therefore, their application in control and decision-support systems may produce a significant beneficial effect.

To preserve the quality of forecasts under the nonstationarity of the process under study and to improve the quality of forecast estimates for processes with arbitrary statistical characteristics, adaptive forecast estimation schemes are applied. The input data for the analysis of forecast quality and for the formation of adaptive schemes of their estimation are the values of forecast errors and their statistical characteristics (variance, standard deviation, and mean values).

P. I. Bidyuk and his disciples proposed methods to adapt forecast estimation schemes to statistical data characteristics in order to improve the quality of forecast estimates. Which scheme is applied depends on specific problem formulation, the quality and amount of experimental (statistical) data, formulated requirements to the quality of forecast estimates and the time interval for the computations. Each method of the adaptive formation of a forecast estimate has specific features, which are taken into account in creating an adaptive prediction system [12].

An error feedback in an adaptive prediction system improves the quality of the model to a level necessary for a high-quality forecast. Error feedback also improves the accuracy of forecast estimates due to enhancing the quality (informativeness) of data and refining the structure of the model. It allows avoiding overlearning, which improves data approximation accuracy but deteriorates the quality of forecast estimates.

A technique was developed to construct a Bayesian network as a directed acyclic graph, intended to model and visualize the information on a specific problem of network training based on available information and to make statistical inference. A Bayesian network is a model that represents probabilistic interrelations among nodes (variables of the process). Formally, it is a triple $N = \langle V, G, J \rangle$ whose first component is a set of variables V, the second is a directed acyclic graph G whose nodes correspond to random variables of the process being modeled, and J is the general distribution of the probabilities of variables $V = \{X_1, X_2, ..., X_n\}$. The set of variables obeys the Markovian condition, that is, each variable of the network depends only on the parents of this variable.

First, the problem of computing the values of mutual information among all nodes (variables) of the network is formulated. Then it is necessary to generate the optimal network structure based on the chosen optimization criterion, which is analyzed and updated at each iteration of the training algorithm.

Given a sequence of n observations $x^n = (d_1, ..., d_n)$, the function of formation of the structure $g \in G$ has the form:

$$\log\left(P(g, x^n)\right) = \log\left(P(g) \cdot \prod_{j \in J}\left(\prod_{s \in S(j,g)}\frac{(\alpha^{(j)} - 1)! \cdot \prod_{q \in A^{(j)}}(n[q, s, j, g]!)}{(n[s, j, g] + \alpha^{(j)} - 1)!}\right)\right)$$

$$= \log\left(P(g)\right) + \sum_{j \in J}\left(\sum_{s \in S(j,g)}\left(\sum_{i=1}^{\alpha^{(j)}-1} i + \sum_{q \in A^{(j)}}\left(\sum_{i=1}^{n[q,s,j,g]} i\right) - \sum_{i=1}^{n[s,j,g]+\alpha^{(j)}-1} i\right)\right)$$

$$= \log\left(P(g)\right) + \sum_{j \in J}\left(\sum_{s \in S(j,g)}\left(\sum_{q \in A^{(j)}}\left(\sum_{i=1}^{n[q,s,j,g]} i\right) - \sum_{i=\alpha^{(j)}}^{n[s,j,g]+\alpha^{(j)}-1} i\right)\right),$$

where $P(g)$ is a priori probability of the structure $g \in G$; $j \in J = \{1, ..., N\}$ denotes the enumeration of the nodes of the structure of the network g; $s \in S(j, g)$ denotes the enumeration of the set of all sets of values taken by the parent nodes of the jth node:

$$n(s, j, g) = \sum_{i=1}^{n} I(\pi_i^{(j)} = s), \quad n[q, s, j, g] = \sum_{i=1}^{n} I(x_i = q, \pi_i^{(j)} = s),$$

where $\pi^{(j)} = \Pi^{(j)}$, $X^{(k)} = x^{(k)}, \forall k \in \phi^{(j)}$, and the function $I(E) = 1$ if the predicate $E = true$, otherwise $I(E) = 0$.

The created network is used as a basis to form a probabilistic inference with the use of training data (two-step procedure).

Step 1. On the set of training data, compute the matrix of empirical values of the general probability distribution of the whole network $P(X^{(1)}, ..., X^{(N)})$:

$$P_{\text{matrix}}(X^{(1)} = x^{(1)}, ..., X^{(N)} = x^{(N)}) = \frac{n[X^{(1)} = x^{(1)}, ..., X^{(N)} = x^{(N)}]}{n},$$

where n is the number of training observations, $x^{(j)} \in A^{(j)}$;

Step 2. Sequentially analyze the state of nodes of the Bayesian network. If a node is uninstantiated, then compute the probabilities of all its possible states. If the values of the nodes of a row coincide with the values of uninstantiated nodes and the state of the node being analyzed, then add the corresponding value to the value of the probability of the corresponding state of the node. Then normalize the probabilities of states of the node being analyzed.

Academicians I. V. Sergienko, M. Z. Zgurovsky (at the center), and Professor V. P. Shilo (second from the left) among students of US universities who underwent practical training in Kyiv

The models were used to perform short- and long-term predictions of the dynamics of financial (bank) and industrial enterprises, to form a strategy of the development of small and medium business using Bayesian networks, to estimate and predict the credit status of clients of a bank, to analyze trends in indices when performing commercial operations at stock exchange, to form the optimal portfolio of securities, and the method of analysis and prediction of risks of portfolio investments.

The IASA participates in the international programs TACIS, EDNEC, INTAS, etc. The achievements of the institute include applied research in socioeconomic problems of the state, common with the IIASA (Austria). The studies in planning and creating a national computer network of educational institutions and sciences (URAN system) are performed under active participation of the institute and are supported by the NATO technical committee. The Research Institute "INFORES" participated in projects of the European Commission CALMAPTACIS and Copernicus-INCO, projects of the Ukrainian Science and Technology Center

(STCU), in the joint project of the Intel company and the Ukrainian and South Korean company SeoduLogic, etc. The state scientific program "Technological forecast as a system methodology of the innovation-based development of Ukraine" was initiated by the institute and was carried out under active cooperation with UNIDO (Austria).

The UNESCO department "Higher education in the sciences, applied systems analysis, and computer science," the World Data Center "Geoinformatics and sustainable development," a base of the UNESCO Institute in information technologies in education, the national committee of the international organization "CODATA," the national committee in information and data of the International Council for Science (ICSU) work under the aegis of the institute.

The IASA cooperates with scientific institutes of the branches of computer science, geosciences, physics and astronomy of the NAS of Ukraine and with higher educational institutions and takes part in solving problems of Kyiv and AR of Crimea.

These data predetermine the urgency and practical necessity of training experts who are able to solve complex system problems of timely prediction, fair forecast and systems analysis both of available socioeconomic, scientific and technical, and other problems, tasks, and situations and of possible technogenic, ecological, natural, and other disasters. Noteworthy is that in practice, much of the efficiency and reliability of timely prediction, objective forecast, systems analysis of different alternatives of complex solutions and strategies depend on how the system researcher can learn in due time and rationally use the capabilities of the methodology of systems analysis.

Fundamental and applied studies at the IASA are carried out in close relation with the educational process. The results of scientific developments are incorporated directly into the educational process. The academic training of students is combined with their practical activities in scientific subdivisions of the institute.

The doctorate and postgraduate studies of the institute are open for researchers in physics and mathematics and in engineering sciences with a major in "Systems analysis and optimal decision theory," "Mathematical modeling and computational methods," "Control systems and processes," "Information technologies," "Systems of automation of design processes," etc.

The target training of senior students involves their obligatory participation in scientific research at the institute and at the place of future work. As experience shows, the graduates successfully work as systems analysts, systems engineers, mathematicians, programmers, LAN administrators, economists at the institutions of the NAS of Ukraine, state authorities of different levels, branch research institutes, and commercial firms and banks.

The up-to-date computer laboratories provide favorable conditions for practical training such as training courses of CISCO network academy in design, construction, and administration of local and global networks, and in the fields related to design of modern integrated circuits and analytic support of banking.

There are faculties of system studies, preuniversity training, course training, second higher and postgraduate education. Experienced scientists and educators from the NTUU "KPI," V. M. Glushkov Institute of Cybernetics, Institute for Information Recording of the NAS of Ukraine, Institute of Mathematics of the NAS of Ukraine, and Institute of Space Research of the NAS of Ukraine and National Space Agency of Ukraine are invited for lecturing, practical training, and supervising thesis works.

Combining the educational process with research studies where senior students take part yields positive results: Foreign universities have become interested in this experience. Not in vain, senior students from the USA who specialize in developing computer-aided technologies win annual competition to go to Kyiv to the Institute of Cybernetics and Institute of Applied Systems Analysis for practical training during which common scientific seminars and conferences are held.

Chapter 3
Mathematical Modeling and Analysis of Complex Processes on Supercomputer Systems

Abstract Optimization methods and problems of control over complicated processes are investigated, and a formalized description of important processes connected with mathematical modeling of complicated environmental phenomena, in particular, phenomena that require the investigation of soil media is given. An approach is proposed to the solution (on supercomputers) of complicated problems that arise in this case. This class of problems is of great importance in the solution of various environmental problems. This problematics is also of interest to geologists, builders, mine builders, and many other specialists. Interest in it has notably increased after the Chernobyl disaster whose intelligent treatment was performed by cyberneticists under the direction of V. S. Mikhalevich. This chapter also considers the optimization of calculations and information security and problems of development of computer technologies oriented toward the investigation of biological processes. It is emphasized that problems that arise in these important lines of investigations are extremely complicated and can be successfully solved only on modern supercomputers.

3.1 Optimization Methods and Control of Complex Processes

In this section, we will dwell on some scientific advances of V. M. Kuntsevich, B. N. Pshenichyi, and their disciples. The studies and the results obtained under the guidance of Pshenichnyi by representatives of his school (and certainly himself) are significant and recognized worldwide. These fundamental and applied studies were conducted at the Institute of Cybernetics till 1996 and then also at the Institute of Applied Systems Analysis and was guided by A. O. Chikrii at the Institute of Cybernetics. The studies were closely related to other studies on new optimization methods and were constantly supported by Mikhalevich.

Pshenichnyi was a representative of so-called traditional optimization and developed theoretical fundamentals of optimization – necessary extremum conditions – by studying the properties of multivalued mappings and convex structures.

I.V. Sergienko, *Methods of Optimization and Systems Analysis for Problems of Transcomputational Complexity*, Springer Optimization and Its Applications 72, DOI 10.1007/978-1-4614-4211-0_3, © Springer Science+Business Media New York 2012

The leaders in this field, A. V. Kantorovich, V. F. Demyanov, O. M. Rubinov, O. Ya. Dubovitskii, A. A. Milyutin, O. D. Ioffee, and V. M. Tikhomirov from Russia, and western scientists, G. Dantzig, S. Karlin, T. Rockafellar, H. Halkin, and L. Neustadt, have raised the research to an unprecedented mathematical level. In this constellation of scientists, Pshenichnyi takes an important place as an expert in the field of necessary extremum conditions and numerical optimization methods.

Pshenichnyi managed to obtain results on convex programming in Banach spaces and on duality in extreme problems, which has brought worldwide recognition to the young author [143]. Convex programming is the most complete part of mathematical programming. The basic results in this field are those due to H. W. Kuhn and A. W. Tucker. Pshenichnyi has developed a general approach covering many classical results. In particular, his technique and the Kuhn–Tucker theorem yield the necessary and sufficient conditions for the solution of a convex programming problem in differential form. He studied the properties of sets of support functionals and related them to directional derivatives. He introduced the class of quasidifferentiable functionals in contrast to convex functionals. This class is rather wide. It includes, in particular, all Gateaux differentiable and convex functionals.

The concept of a quasidifferentiable function allowed describing a function that behaves locally like a convex function. It is a function whose derivative at a point x in any direction exists, is finite, and is a positive homogeneous, convex, and closed function. Then this directional derivative is a support function of some closed convex set and is interpreted as a subdifferential $\partial f(x)$ at the point x of the input function f. Such a definition, on the one hand, substantially expands the class of functions under study and, on the other hand, allows studying them with a technique similar to that applied for convex functions. In particular, this made it possible to reformulate the necessary extremum conditions in mathematical programming problems for nonconvex functions. For example, in the simple case of unconditional optimization, the necessary extremum conditions for a quasidifferentiable function $f(x)$ at a point x_0 remain classical: $0 \in \partial f(x_0)$. The definition of subdifferential for a quasidifferentiable function and its description for various operations with functions was important in developing numerical optimization methods based on subgradient descent.

Quasidifferentiable functionals are also important since the operation of maximization over a parameter keeps the result in the class of quasidifferentiable functionals.

The technique was successfully applied to derive necessary extremum conditions in general mathematical programming problems that generalize the Lagrange multiplier rule to a wide class of problems. In this area, his results are closely related to the Milyutin–Dubovitskii theory, where the necessary extremum conditions for general problems were probably formulated for the first time, and to the studies by L. Neustadt and H. Halkin.

Problems of Chebyshev approximation, optimal control with phase constraints, and discrete principle of maximum were considered from unified standpoints; minimax theorems were proved; criteria to solve a system of convex inequalities were obtained and related to the problem of moments.

As to the duality problem, Pshenichnyi proved the theorem that a convex programming problem can be associated with another convex programming problem with a more complex objective function but simple constraints and whose solution is equivalent to the solution of this problem. As is known from the classical linear programming theory, it is sometimes more convenient to solve the dual problem than the direct one.

Noteworthy is also the following important feature of the dual problem: It is finite dimensional even if the direct problem is not. It is this fact that is widely used in some algorithms to solve linear optimal control problems.

Important concepts such as directional derivative, subgradient, and subdifferential of a convex function were an efficient tool in scientist's hands and allowed attaining a goal when others failed. These concepts are well known today. However, they help to trace the line of his reasoning to introduce the concept of the upper convex approximation of a function, which is one of the major concepts underlying the necessary extremum conditions for nonsmooth and nonconvex functions.

It is obvious that to write necessary extremum conditions, certain conditions should be imposed on the functions that appear in the problem. In particular, near the point of minimum, they should admit the approximation by some simpler functions that can be calculated with relative ease. For example, smooth functions allow linear approximation; convex functions are well approximated by convex positive homogeneous functions – directional derivatives. But nonsmooth and nonconvex functions cannot be approximated by convex positive homogeneous functions near some point. The concept of the upper convex approximation is introduced for such functions. Note that to obtain substantial extremum conditions, a rather wide family of upper convex approximations should be known. The concept of the main upper convex approximation was also introduced; it characterizes how well the behavior of the function near a point can be described by some positive homogeneous function [141].

The general necessary conditions expressed in terms of the upper convex approximations of the function, cones of tangent directions, and Boltyanskii's local tents allowed formulating the necessary extremum conditions in nonsmooth and nonconvex conditional optimization problems under equality and inequality constraints.

The theory of multivalued mappings and its application was one of the areas of Pshenichnyi's scientific interests. Multivalued mappings allow looking at extremum problems from a unified point of view, deepening the understanding of processes and often simplifying the proof of statements. The important concept of a locally conjugate mapping was introduced as well; it is ideologically close to the concept of derivative of a multivalued mapping.

The concepts introduced made it possible to formulate the duality theorem for convex multivalued mappings, from which many statements of convex analysis and theory of extreme problems follow such as the Fenchel–Moreau theorem (quality of a convex closed eigenfunction to a twice conjugate function) and the minimax theorem.

Problems of mathematical economics were an important and topical field of Pshenichyi's interests. Together with Glushkov, he developed an economic dynamics model that takes into account technological progress, variations in technologies, and new technologies. The model takes into account two branches (production and consumption) and limited manpower. Dynamic models were analyzed such as part exchange on a finite time interval, in the presence of a home market. If priority functions are linear, such an exchange model establishes that the equilibrium prices are unique and the function of redundant demand is monotonic, which makes it possible to find the equilibrium prices.

One of the most interesting results in the field of numerical optimization methods [145] is the linearization method [142], which can be applied to solve linear and nonlinear programming problems, systems of equalities and inequalities, and minimax and other problems. The method was published in 1970, and the author developed and improved it all the time. Let us briefly describe the linearization method for a minimization problem with inequality constraints:

$$\min f(x), \quad x \in R^n, \quad g_i(x) \leq 0, \quad i = 1, \ldots, t.$$

One of the modifications of the method can be described as follows. Construct a sequence $x_{k+1} = x_k + \alpha_k p_k$, where α_k is a step factor and p_k is a vector defined as a solution of the quadratic programming problem

$$\min_p \left\{ (f'(x_k), \, p) + \frac{1}{2} \|p\|^2 \right\}, \quad g_i(x_k) + (g'_i(x_k), \, p) \leq 0, \quad i = 1, \ldots, t.$$

The factor $\alpha_k = (1/2)^{j_0}$, where j_0 is the first value of the index $j = 0, 1, \ldots$ that satisfies the inequality

$$f(x_k(1/2)^j p_k) + NF(x_k + (1/2)^j p_k) - f(x_k - NF(x_k)) \leq (1/2)^j \varepsilon \|p_k\|^2,$$

$$0 < \varepsilon < 1.$$

N is a sufficiently large constant and $f(x) = \max \{0, g_1(x), \ldots, g_t(x)\}$.

The implementation of the linearization method implies that the problem solution exists for any x. The function $\Phi_N(x) = f(x) + NF(x)$ plays a special role, both ideologically and numerically. It turns out that for many nonlinear programming problems, $\Phi_N(x)$ is an exact penalty function for sufficiently large N. The problem solution at nonstationary points determines the descent direction of the function $\Phi_N(x)$. The range of nonlinear programming problems for which the conditions whereby $\Phi_N(x)$ is an exact penalty function are satisfied is rather wide, which provides ample opportunities for applying the linearization method.

Supporting optimization research, Mikhalevich did not forget about more complex problems such as control problems where optimization methods play a key role. Large groups of experts in automatic control theory (A. G. Ivakhnenko,

A. I. Kukhtenko, V. M. Kuntsevich, Yu. I. Samoilenko, V. I. Ivanenko) and in optimal control theory (B. N. Pshenichnyi, A. A. Chikrii) worked at the Institute of Cybernetics.

The investigations by the above authors, in particular Kuntsevich, were concerned with the development and dynamic analysis of sampled-data extremum self-adjusting automatic control systems [83]. The wide use of impulsive (discrete-time) elements and means of automation and computing technique stimulated the development of methods for the mathematical description and analysis of nonlinear sampled-data systems with frequency and frequency-width modulation [87]. Kuntsevich and his disciples paid much attention to the analysis of nonlinear difference equations that describe the dynamics of such systems, to the analysis of periodic processes, and to the application of the Lyapunov function method to derive the asymptotic stability and dissipativity conditions. Frequency methods were also developed for the analysis of nonlinear nonstationary control systems, both deterministic and stochastic.

The synthesis of feedback control for deterministic continuous- and discrete-time, generally nonlinear and nonstationary systems, was put in the forefront. However, as is often the case in solving scientific problems, analysis and synthesis problems appeared to be closely related to each other. Synthesis problems were solved based on a unified methodological basis, with the use of Lyapunov functions and their discrete analogues [86]. Synthesis problems for asymptotically stable control systems were solved for a rather wide class of linear and nonlinear control objects under various constraints imposed on the control. The solution of a strengthened problem was considered, where the synthesized control should not only provide the asymptotic stability (or dissipativity) of the system but also minimize some loss function. The solution of the generalized synthesis problem was obtained, according to which the synthesized control should minimize a performance functional. General theorems on the properties of optimal control were proved with the use of Lyapunov functions [85, 86]. For a nonlinear object and general form of the performance functional, a theorem on the optimality "in the small" was proved, that is, on the optimality for small deviations in the nonlinear system of a linear control algorithm, optimal for a system of "linear–quadratic approximation." It was used to develop numerical methods to determine the optimal control for nonlinear discrete systems.

Kuntsevich developed methods to determine suboptimal control algorithms for objects subject to bounded additive external perturbations. For permanent bounded multiplicative perturbations, theorems that allow Lyapunov functions to be used to determine the optimal control were proved. A specific form of an optimal algorithm can be found as the solution of a minimax problem, where the minimum is taken over the control and maximum over the unknown bounded values of external perturbations. In the special case of a linear object with a scalar multiplicative perturbation acting directly on the output of the control unit, an analytic solution of the problem was obtained, which needs only the solution of a matrix equation being a generalization of the well-known Riccati matrix equation. This allowed proving an analogue of the theorem of optimality "in the small" for the nonlinear case.

A class of systems described by matrix difference equations was specified, and original methods of the analysis and synthesis [85] that use direct (or Kronecker) product of matrices were developed for these systems.

A significant contribution was made to the solution of the most topical problem of modern control theory, that is, the analysis and synthesis of control systems under nonstochastic uncertainty, where the values of parameters of the control object are unknown and only the sets in the space of parameters to which they belong are known, and there are similar assumptions on the values of uncontrolled external perturbations. A technique was proposed to obtain guaranteed estimates of object parameters as polyhedrons, which can be obtained directly by set intersection [208].

Kuntsevich initiated and guided the development of a technique for set-valued estimation of the phase state and parameters of an object based on a sequence of approximating ellipsoids. To characterize uncontrolled perturbations, set-valued estimates of their values were used. As a result, rather general mathematical model of an uncertain process (similar to that used in the correlation theory of random processes) was set up, and a technique was proposed to use it for problems of set-membership identification and filtration.

The fundamentals of the theory of adaptive and robust control systems under nonstochastic uncertainty were developed, the optimal control being found from the minimax condition for a given performance criterion (minimum over control and maximum over all the uncertain factors). Since an important uncertainty often remains for the synthesized system after choosing the structure and parameters of the control algorithm, linear and nonlinear systems were analyzed for robust stability [84].

In the case of general nonlinear dynamic systems and additive bounded perturbations, one cannot speak about equilibrium points and their stability. This encouraged the interest in the problem of determining invariant and stationary sets of nonlinear discrete systems under bounded perturbations, and Kuntsevich and Pshenichnyi obtained important theoretical results in this field.

Classical methods of control are described in the monograph by B. M. Bublik and M. F. Kirichenko, who also proposed a number of new approaches to solve the problem of control of dynamic systems described by ordinary differential equations.

The finite-difference method of the approximate solution of optimal control problems is presented in the monograph by Yu. M. Ermoliev, V. P. Gulenko, and T. I. Tsarenko. Their ideas were developed later as applied to stochastic, nondifferentiable, and integer control problems. The results were obtained for the numerical implementation of Bellman strategies, numerical implementation of the Mikhalevich method of sequential analysis of variants, the stochastic gradient method, and Pontryagin's maximum principle.

The problem of minimax estimation of parameters of dynamic systems has been a subject of research by scientists of the Institute of Cybernetics for a long time. A number of fundamental results have currently been obtained in various fields in this subject (M. F. Kirichenko, B. N. Pshenichnyi, V. G. Pokotilo, and G. M. Bakan).

Control of distributed-parameter systems is one of the most difficult problems in control theory. A large scope of studies (theoretical and experimental) was related to maintaining the equilibrium of torus-shaped plasma in a tokamak with automatically controlled transverse magnetic field created by special controlled electromagnets.

These results significantly affected the development of the thermonuclear research program for tokamaks by providing the possibility of holding the plasma in equilibrium. Methods were developed to control plasma evolution in thermonuclear devices, and new approaches were proposed to control the processes in plasma, which are based on models with incomplete information.

Investigations in the theory of control of high-temperature plasma stimulated the development of the general principles and methods to control fast processes in continua. The fundamentals of control systems that correspond to different hierarchy levels of space-time scales of plasma processes were developed.

It turned out to be efficient to use the impedance approach to the analysis and synthesis of a control system that considers the operator of a distributed controlled device as an impedance operator relation between the electric and magnetic components of electromagnetic field at some object-controller interface.

Yu. P. Ladikov-Roev investigated complex dynamic objects and considered modeling of processes, instability analysis, and construction of automatic control systems for objects described by models of kinetics and mathematical physics. Systems of automatic stabilization of the temperature of plasmatron arc and the temperature of thermonuclear burn in a tokamak reactor were designed, and various types of instability in low- and high-temperature plasma, in rare metals, and in fluid flows for large Reynolds numbers were analyzed.

M. F. Kirichenko considered general motion-stabilization problems.

The studies by D. V. Lebedev and A. I. Tkachenko dealt with the theoretical fundamentals of the synthesis of control systems for mobile objects with the information part as a precise and advanced navigator – a platformless inertial navigation system. They presented a method to synthesize control algorithms under constraints for control parameters whose number does not exceed the number of degrees of freedom of the object as a rigid body. They outlined a method to construct algorithms of finding motion parameters based on the data of sensitive elements of the navigation system and obtained highly accurate and economic algorithms to integrate equations. They also examined the possibilities of correcting the platformless navigation system both in the general formulation and with the use of specific sources of extraneous information, including satellite radio-navigation system.

V. V. Pavlov investigated decision making in ergatic control systems, including conflict situations. The developed methods are extended to the class of logic-dynamic systems.

The popular deductive approach needs a great volume of a priori information about the set of variables, constraints on their increase, and basic function of the model. Much less a priori information is necessary for inductive approach, where the optimal model can be found by comparing a large number of candidate models based on an external criterion. The inductive method of modeling and predicting random processes is called the group method of data handling [74].

It is very important to establish constructive relations among general mathematical theories and mathematical problems in the control theory. In this field, it is important to attempt to apply methods of the theory of differential manifolds and fibers, the theory of continuous groups and algebras, and the theory of external differential forms to study controlled dynamic systems.

This has allowed deriving some important results, which is a weighty contribution to the differential-geometric control theory and abstract systems theory.

The theory of decision making related to control of dynamic systems is an important research field. The statistical regularities of random (in a wide sense) phenomena were revealed: Existence theorems for statistical regularities and theorems of dependence of such regularities on solution choice criteria were proved, the measures of uncertainty of random phenomena were investigated and related to objective functions, and in particular, a theorem on unique dependence in Bayesian problems was proved.

The controls of distributed-parameter systems and of solutions of the Fokker–Planck–Kolmogorov equation for probability distributions were analyzed, and the results were obtained for controlled random processes and processes with delay.

Optimal control problems were analyzed for nonlinear, elliptic, parabolic, and hyperbolic systems with constraints in the form of operator inequalities and inclusions imposed on the control and phase variables. In particular, the solvability conditions for control problems for evolution equations and inclusions in Banach spaces were obtained, necessary control optimality conditions in the form of variational inequalities were substantiated, and regularizing approximations of these problems and optimality conditions for them were constructed. The developed variational methods were used to propose a new method of the synthesis of controls in nonlinear distributed systems [73].

For mixed linear systems that contain both units with distributed and lumped parameters, new efficient methods for the stability analysis and synthesis of stabilizing controllers and procedures of their numerical implementations were developed.

The theoretical results were widely used for creating process control systems.

The studies in optimization and control at the Institute of Cybernetics were closely related to leading scientific schools.

The "Stochastic Optimization" international conference was held in Kyiv in 1984; it was organized by the Institute of Cybernetics and Institute for Applied Systems Analysis (Vienna, Austria) mainly due to Yu. M. Ermoliev's efforts. Academicians L. S. Pontryagin and N. N. Krasovskii delivered plenary reports on the Soviet side and T. Rockafellar and G. Vetts from the USA. There was a tightness in the air because of the presence of Americans. During the report of Pontryagin, who lost his sight when he was a teenager, his assistant wrote formulas for him on the board. At some moment, the assistant forgot a denominator in one of the formulas and applied for assistance to Pontryagin. Everybody stood motionless.

Pontryagin was 76 at that time. He calmly dictated the mathematical expression occupied half of the board. The hall erupted in tumultuous applause. By the way, at the international conference dedicated to the 100th anniversary of L. S. Pontryagin (2008) in Moscow, he was recognized as one of the world's best mathematicians of the second half of the twentieth century.

Academician M. M. Krasovskii from Sverdlovsk (now Ekaterinburg) was the second leader on the Soviet side at this conference. In Kyiv, he was accompanied by Prof. A. A. Chikrii whom Krasovskii opposed a few years ago in defending the doctoral dissertation. The conference was held in early September, and September 7, 1984, was the 60th birthday of Krasovskii. Naturally, it would be desirable to arrange a celebration for the guest; at the same time, Krasovskii was known to disallow ceremonies. Mikhalevich invited Krasovskii to the Institute of Cybernetics for excursion, showed him the departments, and detailed the story of scientific achievements. At the pattern recognition department, the story touched upon the Bellman dynamic programming method. Mikhalevich and Krasovskii were delighted and, with chalk in their hands, discussed all the mathematical details of this method for a long time. It seemed to never end .... Finally, Academician Krasovskii was congratulated on his anniversary, easily and friendly, as it was usual in Kyiv Teremky.

The theory of differential games or conflict-controlled processes is an important link of the modern mathematical control theory.

Along with world-renowned schools, Pontryagin's in Moscow and Krasovskii's in Ekaterinburg, the Kyiv School has been successful for 40 years. It was headed by B. N. Pshenychnyi, and now A. A. Chikrii continues his work. By the way, when Chikrii became a head of a separate laboratory, he proposed to call it the laboratory of conflict-controlled processes. Mikhalevich read this title and, after a thought, told that there are so many conflicts in our life that it would be better to do without another one in the title .... The laboratory, and later the department of optimization of controlled processes, still exists at the Institute of Cybernetics.

Fundamental methods to solve dynamic game problems were developed during these years. In particular, the concept of ε-strategy was introduced and the game of optimal pursuit time in the class of these strategies was shown to always have a saddle point. A key point in this theory is constructing a semigroup of operators that map a set into a set and proving that given this semigroup, the problem can be solved completely [146].

Operator structures and ε-strategies are convenient tools to describe the structure of a game. Moreover, ε-strategies are equivalent to strategies where the pursuer uses only information on the initial position and control history of the opposite party and are equivalent to position strategies under additional assumptions. This allows implementing ε-strategies in specific problems. For example, for a wide class of linear problems, the above operators can be set up quite efficiently and the calculations be carried out explicitly. Using methods of convex analysis, including H-convex and matrix-convex sets, plays a decisive role.

An important approach to solving game problems is replacing a continuous-time game with a discrete-time one. Approximate methods were developed to solve differential games, such as the Euler method for differential equations. The difference of exact and approximate solutions was estimated.

Convex analysis and mathematical programming methods such as directional differentiation of functions of maximum were used, and sufficient conditions were obtained for the termination of a differential game in absorption and maximum time.

The solution of an optimal control problem as an extremum problem for differential inclusions was considered. A technique was developed to study special multivalued mappings (convex and convex-valued). The concept of a locally conjugate multivalued mapping was introduced and appeared an efficient tool to study complex extremum problems.

A qualitative theory of linear and nonlinear discrete conflict-controlled processes was developed. Foundations were laid for the nonlinear theory of collision avoidance in dynamic game problems; in particular, methods of directional evasion, variable directions, invariant subspace method, and recursive method have been developed; and the Pontryagin–Mishchenko maneuver-detour method has been generalized.

Sufficient conditions of the first and higher orders in fine and rough cases were obtained, the conditions for evading a group of pursuers were established, and interaction under conflict was analyzed. In particular, Pontryagin's μ-problem was solved.

A formula was derived to represent the solutions of the nonlinear conflict-controlled system:

$$\dot{z} = f(z, u, v), \quad u \in U, \quad v \in V,$$

where M_0 (a linear subspace) is a terminal set, $L = M_0^\perp, W \subset L, \quad \pi_W$ is an orthoprojector $\pi_W : R \to W$, and

$$\varphi^{(i)}(z, u, v) = (\nabla_z \varphi^{(i-1)}(z, u, v), f(z, u, v)), \quad \varphi^0(z, u, v)) = \pi_W z, \quad i = \overline{1, k},$$

provided that $\varphi^i(z, U, V) = \varphi^{(i)}(z), i = \overline{1, k-1}$. It is called the Pshenichnyi–Chikrii formula:

$$\pi_W z(t) = \sum_{i=1}^{k-1} \frac{t^i}{i!} \varphi^{(i)}(z_0) + \int_0^t \frac{(t-\tau)^{k-1}}{(k-1)!} \varphi^{(k)}(z(\tau), u(\tau), v(\tau))d\tau.$$

This formula became the basis for the development and analysis of sufficient evasion conditions. In the case of a group of pursuers, an important function was introduced in the evasion problem:

$$\omega(n, v) = \min_{||p_i||=1} \max_{||v||=1} \min_{i=\overline{1, v}} |(p_i, v)|, \quad p_i, v \in R^n, \quad n, v \in N, \quad n \geq 2, \quad v = 2.$$

The exact values of this function are known only when $n = 2$ or $v = 2$, in other cases, the lower bounds are known. This function made it possible to formalize the situation of surrounding.

The evasion problem covers a practically important collision avoidance problem for aircraft at landing and takeoff, for ships in and nearby seaports, in channels, etc.

The developed methods allow supervising mobile objects so as to provide accident-free traffic in space and in ports, to help a dispatcher to promptly influence the situation, and to assign air and sea safety corridors. Such a computer system was developed by V. V. Pavlov and his colleagues.

A position method of group and serial pursuit that generalizes the Krasovskii rule of extremum aiming was proposed. The cases of constraints on the state and of information delay were examined, and sufficient conditions for the complete conflict controllability of quasilinear systems were established. In particular, it was established that for simple motion with spherical domains of control, the locus of catching of the evader is the limacon of Pascal.

The method of resolving functions, a new method in the pursuit theory, which is based on inverse Minkowski functionals and multivalued mappings, was developed [177, 197]. The Minkowski and Fenchel–Moreau theory of duality to study terminal game problems with the use of conjugate functions and polar sets was put into use.

Since the method is important, we will dwell in more detail on it. Let us consider a quasilinear conflict-controlled process

$$\dot{z} = Az + \varphi(u, v), \quad z \in R^n, \ u \in U, \ v \in V, \ U, V \in K(R^n),$$

with cylindrical terminal set

$$M^* = M_0 + M,$$

where M_0 is a linear subspace, $M \in K(L)$, $L = M_0^\perp$, and $\varphi(u, v)$ is a function continuous in the totality of variables.

The pursuer chooses the control as

$$u(t) = u(z_0, v_t(\cdot)), \quad v_t(\cdot) = \{v(s) : s \in [0, t]\}.$$

The task is to steer the trajectory of the process to the set M^* for any action of the evader.

Let π be an orthoprojector from R^n onto L. Let us consider multivalued mappings

$$W(t, v) = \pi e^{At} \varphi(U, v), \quad W(t) = \bigcap_{v \in V} W(t, v).$$

The Pontryagin condition is said to be satisfied if $W(t) \neq \varnothing \ \forall t \geq 0$. Then there exists a measurable selector $\gamma(t) \in W(t)$. Denote

$$\xi(t,\ z,\ \gamma(\cdot)) = \pi e^{At}z + \int\limits_0^t \gamma(\tau)d\tau,$$

and introduce the resolving function by the formula

$$\alpha(t,\ \tau,\ v) = \sup\ \{\alpha \geq 0 : [W(t-\tau,\ v) - \gamma(t-\tau)] \cap \alpha[M - \xi(t,\ z,\ \gamma(\cdot))] \neq \varnothing\}$$

and the set

$$T(z,\ \gamma(\cdot)) = \left\{t \geq 0 : \int\limits_0^t \inf_{v \in V} \alpha(t,\ \tau,\ v)d\tau \geq 1\right\}.$$

Theorem 3.1 [197]. *Let the Pontryagin condition be satisfied for a quasilinear conflict-controlled process in the initial state z_0 and $M = \mathrm{co}\ M$. Then, if there exist a measurable selector $\gamma(t)$, $\gamma(t) \in W(t)$, $t \geq 0$, such that $T(z_0,\ \gamma(\cdot)) \neq 0$ and a finite number $T \in T(z_0,\ \gamma(\cdot))$, then the trajectory of the process can be brought to the set M^* from the initial state z_0 in the time T.*

The method theoretically substantiates the classical rule of parallel pursuit for various game situations. Along with the basic scheme of the method, its modifications were developed; pursuit time was compared, in particular, with the time of the first direct Pontryagin method and the time of the first Krasovskii absorption. The time was optimized with respect to Borel selectors, and recommendations were given as to the a priori choice of extremum selectors for a wide range of problems. A differential game was related to a convex problem with the use of sliding mode technique. The scheme of the method applies to the case of constraints on the state and delayed information about the state. Pontryagin's condition is much weakened for oscillating processes, for objects with different lag, and for game problems with non-fixed time. Processes of variable structure were analyzed. The properties of special multivalued mappings related to conflict-controlled process were investigated. The method was applied to solve problems of group and serial pursuit in constructive form.

Problems of takeoff and landing of aircraft under permanent perturbations, in particular "soft landing," are a separate application of studies in the field of conflict-controlled processes. The last problem came from the USA. Its authors are Americans J. Albus and A. Meystel from the National Institute of Standards and Technology (Gaithersburg). The problem is to land an aircraft onto an aircraft carrier, that is, geometrical coordinates and velocities of the objects should coincide. Such a model was developed, and a system that models this process was created.

Bilinear Markov models of search for moving objects under various information assumptions were proposed. To solve the search problem, discrete Pontryagin's maximum principle and Bellman's dynamic programming method were applied.

A search using a group of objects with and without information exchange in the group and hidden search were considered and a search with interaction of groups was analyzed.

Dynamic search models allow efficient mineral exploration, search for accident victims in remote areas, and search for fish shoals and even bugs in programs. The corresponding computer system was once developed by order of Navy by the departments of A. A. Chikrii and V. V. Skopetskii and dealt with searching for and tracking submarines from surface ships, helicopters, and acoustic means.

Processes with pulse control [82], that is, with discontinuous trajectories, were analyzed based on a certain procedure and the method of resolving functions, where classes of strategies are extended to Schwartz distributions.

Pulse control has a direct relationship to control of spacecraft. The developed game methods were applied to solve problems of the interaction of groups of controlled space objects (star wars problem). In particular, target assignment problems, which are of traveling-salesman type, play an important role in this process. Applying the methods of resolving functions and extremum aiming, which substantiate parallel pursuit and pursuit curve, together with the cluster technologies developed at the Institute of Cybernetics, optimizes the interaction of groups due to paralleling. In particular, the developed technologies based on mathematical methods allow successful antiterrorist actions on target (including group) interception.

The developed game methods allow simulating air combat. Such studies were performed in collaboration with the Riga Aviation Institute.

The concepts of conjugate and self-conjugate games were introduced. Game problems for dynamic processes with aftereffect, in particular, differential–difference and integro-differential games, and game problems for systems with fractional derivatives were considered [196].

The monograph [63] deals with the analysis of Nash equilibrium strategies. It proposes a number of new concepts of optimality and relates them with the classical concepts.

3.2 Mathematical Modeling of Processes in Soil

Various complex objects have multicomponent structure and their components differ in mechanical, physical, and other properties. Multicomponent bodies often contain man-made or natural interlayers, thin or short inclusions, etc. Despite their small thickness, such interlayers (inclusions) often have a strong effect on the stress–strain state of and the temperature field in the whole body; and for soil, they influence the motion of fluid in it. Thus, it would be unreasonable to neglect such interlayers (inclusions) in the analysis of the behavior of certain objects. Moreover, such processes in real objects are purely spatial, which should be taken into account.

Many years' experience suggests that the influence of interlayers or cracks on, for example, the stress–strain state of the whole multicomponent body or near the interlayers (cracks) can be taken into account with the help of so-called interface conditions (differential equations referred to the mid-surfaces of these inclusions, which adequately describe the influence of inclusions on the process in question). A special feature of such mathematical models is that they generate classes of boundary-value and initial–boundary-value problems for partial differential equations with discontinuous solutions [30–33, 218], for example, discontinuous tangential displacements (on opposite crack faces or on opposite sides of low-strength interlayers, etc.).

Based on classes of discontinuous functions, classical generalized problems defined on classes of discontinuous functions are obtained for a wide set of classes of such problems. The use of classes of discontinuous functions of finite-element method (FEM) made it possible to set up and theoretically justify highly accurate computational schemes for their numerical discretization [30–33].

In solving problems of mitigating the consequences of the Chernobyl disaster (closely attended to by Mikhalevich), the proposed mathematical models and developed efficient numerical methods were used to analyze the expediency of creating a drainage barrier [33] between the cooling water pond of Unit 4 and the Pripyat River. The problem of freezable bank of the Vilyui River, near the Vilyui hydroelectric power station GES-3, was resolved as well.

A feature of these mathematical models is not only that they form a new class of problems (problems for partial differential equations with discontinuities) but also that they impart new quality to the solutions of these problems. In particular, a computational experiment on the free longitudinal vibrations of a compound rod with low-strength short component revealed cases where some eigenvalues do not depend on this low-strength component [33] while the other eigenvalues decrease as the parameter of the low-strength inclusion decreases. The analytic solution obtained much later by other authors confirmed this fact.

Problems of soil ecology and rational environmental management require numerical analysis of the processes in large multicomponent soil masses, which is usually reduced to the solution (often repeated solution) of systems of algebraic equations with more than 10^5 unknowns. This became possible with supercomputers such as SCIT series.

It should be noted that reliable data on the mechanical and physical properties of components and on external loads are frequently insufficient for successful analysis of processes in real multicomponent soil objects (similar problems also arise in overhaul-period renewal for units of assemblies, etc.). Therefore, it is important to create information technologies that would make it possible to use certain monitoring data for missing input data of mathematical models and to successfully complete calculations.

The following results were obtained: Mathematical models of processes of multicomponent bodies (classes of differential problems with discontinuous solutions), corresponding classes of classical generalized problems defined on sets of discontinuous functions, highly accurate computational algorithms for their numerical discretization, a technique was developed, and explicit expressions for gradients of quadratic residual functionals were derived (based on the results from optimal control theory for states of a set of classes of multicomponent distributed systems) to identify, by gradient methods, different parameters of the set of classes of multicomponent distributed systems and external loads on them. These results form the basis of the *technology of systems analysis of multicomponent objects*.

Let us consider some problems on the stress–strain state (SSS) of multicomponent bodies.

1. Stationary Problems [32, 33]

   The following mathematical model may underlie the analysis of the spatial stress–strain state of a body that contains two components Ω_1, $\Omega_2 \in R^3$ and an interlayer.

   In bounded connected strictly Lipschitz domains Ω_1, $\Omega_2 \in R^3$, a system of elastic equilibrium equations

$$-\sum_{k=1}^{3} \frac{\partial \sigma_{ik}}{\partial x_k} = f_i(x), \quad i = \overline{1, 3}, \tag{3.1}$$

is defined, where $x = (x_1, x_2, x_3)$ is a point of the Cartesian coordinate system; $\sigma_{ki} = \sum_{l,m=1}^{3} c_{iklm} \varepsilon_{lm}$, σ_{ki} and ε_{lm} are elements of stress and strain tensors, respectively; c_{iklm} are elastic constants, $y = (y_1, y_2, y_3)$ is displacement vector, and y_i, $i = \overline{1, 3}$, is its projection onto the ith axis of the Cartesian coordinate system. Mixed boundary conditions are specified on the boundary $\Gamma = (\partial \Omega_1 \cup \partial \Omega_2) \backslash \gamma$, $\gamma = \partial \Omega_1 \cap \partial \Omega_2 \neq \varnothing$:

$$\tilde{l}_1(y) = \varphi_1; \tag{3.2}$$

on the interval γ, which splits the domain $\bar{\Omega}$ into two, $\bar{\Omega}_1$ and $\bar{\Omega}_2$, interface (imperfect bonding) conditions

$$\tilde{l}_2(y) = \varphi_2 \tag{3.3}$$

of the low-strength interlayer, viscous friction, and disjoining pressure [32]. There are inhomogeneous principal conditions among them, for example, if the contacting surfaces are rough, etc.

Each ith problem is established to have a unique generalized solution (that admits discontinuity on γ) as a function $y^i \in V_i$ that minimizes the energy functional

$$\Phi_i(v) = a_i(v,\ v) - 2l_i(v) \tag{3.4}$$

and is a unique function $y^i \in V_i$ that satisfies the following identity $\forall v \in V_i^0$:

$$a_i(y^i, v) = l_i(v). \tag{3.5}$$

Classes of discontinuous FEM functions are used to develop highly accurate computational schemes to solve equivalent problems (3.4) and (3.5). For an approximate FEM solution U_k^{iN} that admits discontinuity on γ, the following estimates hold:

$$\|y^i - U_k^{iN}\|_{a_i} \le ch^k, \quad \|y^i - U_k^{iN}\|_{W_2^1} \le c_1 h^k, \tag{3.6}$$

where k is the degree of FEM polynomials and h is the greatest diameter of all the finite elements. The condition numbers of the FEM matrices are cond $(A) \le ch^{-2}$.

2. Dynamic Problems

The study [31] considers a dynamic problem on the SSS of a body $\Omega = \Omega_1 \cup \Omega_2$, where sliding with friction is possible on a crack, and masses m_1 and m_2 are concentrated on its faces. The mathematical model is based on the variational equation

$$\int_\Omega \sum_{i,j=1}^3 \sigma_{ij} \delta\varepsilon_{ij} dx + \int_\gamma r[u_s]\delta[u_s] d\gamma + \int_\Omega \rho \frac{\partial^2 u}{\partial t^2} \delta u dx + \int_\gamma m \frac{\partial^2 u_n}{\partial t^2} \delta u_n d\gamma +$$

$$+ \int_\gamma m_1 \frac{\partial^2 u_s^-}{\partial t^2} \delta u_s^- d\gamma + \int_\gamma m_2 \frac{\partial^2 u_s^+}{\partial t^2} \delta u_s^+ d\gamma = \int_\Omega f \delta u dx, \tag{3.7}$$

where $m = m_1 + m_2$.

Based on classes of discontinuous FEM functions for approximate generalized solutions $U_k^N(x, t)$, we obtain the estimate

$$\|u - U_k^N(x, t)\| \le ch^k. \tag{3.8}$$

To solve the intermediate Cauchy problem, the Crank–Nicolson difference scheme is used. The estimate

$$\max_{j=0,\ m} \|z_j\| = \|y(\cdot,\ t_j) - U^j(\cdot)\| \le c(h^k + \tau^2) \tag{3.9}$$

holds for the error $z_j(x)$ of the approximate solution obtained with the use of discontinuous FEM functions and Crank–Nicolson schemes, where τ is time step.

3. Eigenvalue Problems for Operators in the Theory of Elasticity

Dynamic problems of the deformation of layered bodies lead to spectral problems [32] for operators of elasticity with discontinuous eigenfunctions. Given surfaces of concentrated masses, problem (3.7) reduces to a spectral problem for an operator of elasticity with eigenvalues in interface conditions.

For such differential spectral problems, classical generalized (weak) spectral problems with eigenfunctions that can have discontinuities on the mid-surfaces of low-strength interlayers are obtained. Classes of discontinuous FEM functions are used to set up computational schemes with highly accurate numerical discretization. For the first n_k approximate numbers λ_i^N and the corresponding approximate eigenfunctions u_{ik}^N, the estimates

$$0 \le \lambda_i^N - \lambda_i \le c_i h^{2k}, \quad \|u_i - U_{ik}^N\|_{W_2^1} \le c_i' h_i^k, \quad i = \overline{1, \, n_k},$$

are obtained.

These estimates indicate that the accuracy of the proposed computational algorithms for spectral problems with discontinuous eigenfunctions is no worse than that for spectral problems with smooth eigenfunctions.

Highly accurate schemes are also set up for thermoelastic problems for multicomponent bodies.

4. Unsteady Motion of a Fluid in Compressible–Incompressible Soil

If the porous component Ω_1 of the body Ω is elastic, then the unsteady motion of the fluid in it can be described by the parabolic equation

$$\beta_0 \frac{\partial u}{\partial t} = \sum_{i,j=1}^{3} \frac{\partial}{\partial x_i} \left(k_{ij} \frac{\partial u}{\partial x_j} \right) + f, \quad (x, \, t) \in \Omega_{1T}, \tag{3.10}$$

where β_0 is elastic capacity factor, k_{ij} are components of the filtration coefficient of the anisotropic body Ω_1, f is the capacity of the source or sink, and u is hydraulic pressure head.

In the incompressible porous component Ω_2 of the body Ω $(\Omega = \Omega_1 \cup \Omega_2)$, the unsteady motion of fluid can be described by the elliptic equation

$$-\sum_{i,j=1}^{3} \frac{\partial}{\partial x_i} \left(k_{ij} \frac{\partial u}{\partial x_j} \right) = f, \quad (x, \, t) \in \Omega_{2T}. \tag{3.11}$$

Mixed boundary conditions are specified on the boundary Γ_T ($\Gamma_T = \Gamma \times (0, T)$, $\Gamma = (\partial\Omega_1 \cup \partial\Omega_2)\backslash\gamma$, $\gamma = \partial\Omega_1 \cap \partial\Omega_2 \neq \emptyset$):

$$u|_{\Gamma_{1T}} = \varphi,$$

$$\sum_{i,j=1}^{3} k_{ij} \frac{\partial u}{\partial x_j} \cos(n, \, x_i) = \beta_1, \quad (x, \, t) \in \Gamma_{2T}, \tag{3.12}$$

$$\sum_{i,j=1}^{3} k_{ij} \frac{\partial u}{\partial x_j} \cos(n, x_i) = -\alpha u + \beta, \quad (x, t) \in \Gamma_{3T},$$

where $\Gamma = \overset{3}{\underset{i=1}{\cup}} \Gamma_i$, $\Gamma_i \cap \Gamma_j = \varnothing$ for $i \neq j$; $i, j = \overline{1, 3}$ and $\Gamma_{iT} = \Gamma_i \times (0, T)$.

On the interval $\gamma_T = \gamma \times (0, T)$, the interface conditions for the weakly permeable interlayer have the form

$$[q] = 0, \quad \{q\}^{\pm} = r[u], \tag{3.13}$$

where $q = \sum_{i,j=1}^{3} k_{ij} \frac{\partial u}{\partial x_j} \cos(n, x_i)$ and n is the normal to γ directed into Ω_2, $r = \mathrm{const} \geq 0$.

For $t = 0$, we have the initial condition

$$u(x, 0) = u_0(x), \quad x \in \bar{\Omega}_1 \cup \bar{\Omega}_2. \tag{3.14}$$

The generalized condition of the initial–boundary-value problem (3.10)–(3.14) is a function $u(x, t) \in W(0, T)$ that satisfies the equality

$$\int_{\Omega_1} \beta_0 \frac{du}{dt} w dx + a(u, w) = (f, w) + (\beta_1, w)_{L_2(\Gamma_2)} + (\beta, w)_{L_2(\Gamma_3)}, \quad t \in (0, T),$$

$$(\beta_0 u, w)_{L_2(\Omega_1)}(0) + (u, w)_{L_2(\Omega_2)} = (\beta_0 u_0, w)_{L_2(\Omega_1)} + (u_0, w)_{L_2(\Omega_2)} \tag{3.15}$$

$\forall w(x) \in V_0$, where $a(u, w) = \int_{\Omega} \sum_{i,j=1}^{3} k_{ij} \frac{\partial u}{\partial x_j} \frac{\partial w}{\partial x_i} dx + \int_{\gamma} r[u][w] d\gamma + \int_{\Gamma_3} \alpha u w d\Gamma_3$.

The solution of problem (3.15) is shown [218] to exist and be unique. For the approximate generalized discontinuous solution $u_k^N(x, t)$ of problem (3.10)–(3.14), the following estimate holds [30]:

$$\|u - u_k^N\|_{H^1 \times L_2} \leq ch^k. \tag{3.16}$$

The intermediate Cauchy problem is solved by the Crank–Nicolson scheme.

For the approximate solution obtained with the use of discontinuous FEM functions and Crank–Nicolson scheme, the estimate [30]

$$\|z_m\|_{1,L_2}^2 + \delta\tau \sum_{j=0}^{m-1} \|z_{j+1/2}\|_{H^1}^2 \leq c(h^{2k} + \tau^4) \tag{3.17}$$

holds, where $\delta = \mathrm{const} > 0$.

5. Unsteady Motion of a Fluid in Block-Fractured Soil

The motion of a fluid in a block-fractured soil can be described by a pseudoparabolic equation. If the domain Ω consists of two bounded open disjoint strictly Lipschitz domains Ω_1 and Ω_2 from R^3 (which correspond to block-fractured components of the medium under study), then we have the pseudoparabolic equation in the domain $\Omega_T = \Omega \times (0,\ T)$ $(\Omega = \Omega_1 \cup \Omega_2)$

$$-\sum_{i,j=1}^{3} \frac{\partial}{\partial x_i}\left(a_{ij} \frac{\partial^2 u}{\partial x_j \partial t}\right) + \bar{a}(x)\frac{\partial u}{\partial t} - \sum_{i,j=1}^{3} \frac{\partial}{\partial x_i}\left(k_{ij}\frac{\partial u}{\partial x_j}\right) = f(x,\ t), \qquad (3.18)$$

where

$$\sum_{i,j=1}^{3} a_{ij}\xi_i\xi_j \geq \alpha_0 \sum_{i=1}^{3} \xi_i^2,\ \sum_{i,j=1}^{3} k_{ij}\xi_i\xi_j \geq \alpha_1 \sum_{i=1}^{3} \xi_i^2,\ \alpha_0,\ \alpha_1 = \text{const} > 0.$$

Mixed boundary conditions are specified on the boundary Γ_T:

$$u = g,\ (x,\ t) \in \Gamma_{1T},$$

$$\sum_{i,j=1}^{3} \left(a_{ij}\frac{\partial^2 u}{\partial x_j \partial t} + k_{ij}\frac{\partial u}{\partial x_j}\right)\cos(n, x_i) = g,\ (x,\ t) \in \Gamma_{2T}, \qquad (3.19)$$

$$\sum_{i,j=1}^{3} \left(a_{ij}\frac{\partial^2 u}{\partial x_j \partial t} + k_{ij}\frac{\partial u}{\partial x_j}\right)\cos(n, x_i) = -\alpha u + \beta,\ (x,\ t) \in \Gamma_{3T}.$$

On $\gamma_T = \gamma \times (0,\ T)$ $(\gamma = \partial\Omega_1 \cap \partial\Omega_2 \neq \varnothing)$, the interface conditions for a three-layer thin interlayer have the form

$$R_1 q^- + R_2 q^+ = [u],\ [q] = \varpi, \qquad (3.20)$$

where γ is a weakly permeable thin interlayer between the domains Ω_1 and Ω_2,

$$q = \sum_{i,j=1}^{3} \left(a_{ij}\frac{\partial^2 u}{\partial x_j \partial t} + k_{ij}\frac{\partial u}{\partial x_j}\right)\cos(n,\ x_i).$$

The initial condition at $t = 0$ is

$$u(x;\ 0) = u_0(x),\ x \in \bar{\Omega}_1 \cup \bar{\Omega}_2. \qquad (3.21)$$

The generalized solution of the initial–boundary-value problem (3.18)–(3.21) is a function $u(x,\ t) \in W(0,\ T)$ that satisfies the following equality $\forall w(x) \in V_0$:

$$a_0\left(\frac{\partial u}{\partial t},\ w\right) + a(u,\ w) = l(w),\ t \in (0,\ T), \qquad (3.22)$$

$$a_0(u,\ w)(0) = a_0(u_0,\ w).$$

The solution of problem (3.22) exists and is unique. Estimate (3.16) holds for the approximate generalized discontinuous FEM solution $u_k^N(x, t)$ of problem (3.18)–(3.21).

The intermediate Cauchy problem can be solved by the Crank–Nicolson scheme. For the error of the approximate solution of problem (3.18)–(3.21) obtained with the use of discontinuous FEM functions and the Crank–Nicolson difference scheme, the estimate [30]

$$\|z_m\|_{a_0}^2 + \eta\tau \sum_{j=0}^{m-1} \|z_{j+1/2}\|_V^2 \le c(h^{2k} + \tau^4)$$

is true, where $\eta, c = \text{const} > 0$, $\tau = T/m$, and m is the number of time steps.

The error estimates of the approximate solutions obtained for the considered classes of mathematical models characteristic of multicomponent soil with thin interlayers, cracks, inclusions, etc., [30–33] certify that they are not worse in the orders of spatial and time steps than those for the classes of problems with smooth solutions.

Similar results are also obtained for conditionally well-posed problems with discontinuous solutions [32], in particular, for the problem on steady filtration of a fluid in a multicomponent medium with thin inclusions/cracks and the Neumann boundary conditions.

6. Information Technologies of Analysis of Processes in Multicomponent Media
   New classes of mathematical differential models with interface conditions are constructed to describe the basic processes in multicomponent soil media with arbitrarily located natural or artificial thin inclusions/cracks; new classical generalized problems defined on classes of discontinuous functions are obtained; a technique is developed to determine the corresponding classes of discontinuous FEM functions; the constructed discrete models and computational algorithms represent a theoretical basis for a *new information technology of studying the basic processes in multicomponent soil objects with various inclusions*.

   These theoretical results were implemented in NADRA and NADRA-3D software systems for the analysis of steady and unsteady motion of a fluid in multicomponent bodies with various arbitrarily located thin inclusions, temperature fields in such objects, and their deformation [31] in two-dimensional and three-dimensional formulations, respectively. The class of mathematical problems for complex spatial objects solved with the use of these information technologies becomes wider.

   To obtain a comprehensible computer solution to a specific problem with significant geometrical dimensions, it is necessary to use finite-element partition with a great number of finite elements. This requires intermediate highly dimensional (over $5 \cdot 10^5$ unknowns) systems of linear and nonlinear algebraic equations to be formed and solved. In this connection, IT NADRA-3D was designed so that intermediate systems of algebraic FEM equations are generated and solved using SCIT supercomputer complexes, and other subsystems of IT NADRA-3D operate on powerful personal computers.

Noteworthy is that the information technology developed for the analysis of processes of multicomponent bodies with inclusions and interlayers can be applied to study not only soils but also multipurpose objects. In particular, under the scientific supervision of one of the authors of this development, at the I. Pulyui Ternopil State Technical University, it was extended to tribology (a science that analyzes the friction of components of multicomponent bodies) problems.

NADRA-3D system was used to analyze the spatial motion of a fluid in a complex soil mass of the Kyiv industrial–urban agglomeration as of 1942 (a year that preceded heavy use of underground waters) and as of the period 1942–1969 (when the city constantly increased the use of underground water).

7. Optimal Control of States of Multicomponent Distributed Systems

In order to preserve soil layers in natural or nearly natural states, it is rather important to preliminarily analyze the expediency of applying certain technologies and their consequences in mining, hillside building, deforestation, building skyscrapers on flat territories, creating various underground communications, etc., which exercises a significant influence on the present and future of people.

Certain measures (such as draining, heat removal, etc.) may improve the state of multicomponent soil. The monographs [30, 31, 218] address the optimal control of the main processes characteristic of multicomponent bodies with various arbitrarily located thin inclusions.

As mentioned above, the main processes characteristic of natural or artificial soil are various flow regimes, formation of temperature states, and deformation. The description of the influence of thin inclusions on the processes under study leads to essentially new classes of mathematical problems, namely, problems with discontinuous solutions, which are mathematical models of states of multicomponent distributed systems.

Mathematically, choosing optimal preliminary measures in such systems reduces to determining the optimal perturbations of boundary conditions, interface conditions, etc., in order to provide the minimum deviation of the state (over the whole body, on its separate areas) from the desirable one. Optimal control of homogeneous distributed systems with quadratic performance functionals $J(v)$ is analyzed in the well-known monograph by J.-L. Lions. The results of this analysis were developed and generalized in the monographs [30, 31, 218] to new important classes of inhomogeneous distributed systems. The states of these systems are described by boundary-value problems for elliptic equations (including conditionally well-posed ones, with nonunique discontinuous solutions), boundary, initial–boundary-value problems for equations of elastic equilibrium of multicomponent bodies with interface conditions, and initial–boundary-value problems for parabolic, elliptic–parabolic, pseudoparabolic, hyperbolic, and pseudohyperbolic systems with various interface conditions, etc.

Optimal control of processes of pressure filtration of a fluid in multicomponent soil and of the formation of steady temperature fields in such bodies reduces
to optimal control problems for states described by elliptic equations with
boundary and interface conditions. The studies [30, 31, 218] consider various
possibilities for optimal control of internal sources, parameters of boundary
conditions, interface conditions, and parameters of different compound multicomponent bodies with various observations of the states of systems (over the
whole multicomponent body, on a part of or its entire boundary, on a thin
inclusion, on the values of the state on different components of the body
simultaneously, etc.). For the considered cases, theorems on the existence of
optimal controls of the states of systems with quadratic cost functionals are
proved, that is, it is proved that there exists a unique element $u \in U_\partial \subset U$ that
satisfied the condition

$$J(u) = \inf_{v \in U_\partial} J(v),$$

where U is the space of controls, U_∂ is the set of feasible controls,

$$J(v) = \|Cy(v) - z_g\|_H^2 + (Nu, \ u)_U,$$

H is a Hilbert space, z_g is a given element of the space H, $z(v) = Cy(v)$ is an
observation, and $y(v) \in V$ is a solution of the problem on the system state for a
given control $v \in U$, $C \in L(V; H)$, $(Nu, \ u)_U \geq v_0\|u\|_U^2$, $v_0 = \text{const} > 0$. The
cases where the boundary-value problems that describe the state include the
principal inhomogeneous boundary conditions and inhomogeneous principal
interface conditions are also considered.

If the set of feasible controls U_∂ coincides with the corresponding complete
Hilbert space U of controls, it is shown that using classes of discontinuous FEM
functions, highly accurate computational schemes can be set up to determine
optimal controls.

The analysis of optimal control of steady diffusion processes for flows specified
on the boundary of a multicomponent body reduces to the analysis of the continuous dependence of a discontinuous state of the system (which can be nonunique)
on various perturbations (controls). It is shown that if a system state $y = y(x, \ u)$
$= y(u)$ is defined among functions of the set $V_Q = \{v \in \bar{H} : \int_\Omega vdx = Q\}$, where
$x \in \bar{\Omega}_1 \cup \bar{\Omega}_2$ and Ω_1 and Ω_2 are the components of a two-component body $\bar{\Omega}$ with
inclusion $\gamma = \partial\Omega_1 \cap \partial\Omega_2 \neq \emptyset$, Q is an arbitrary fixed real number, $\bar{H} = \{v : v$
$|_{\Omega_i} \in W_2^1(\Omega_i), \ i = 1, \ 2\}$, and $u \in U_\partial$ is a control, then the state $y(u)$ and its traces
on certain components of the body $\bar{\Omega}$ continuously depend on various controls,
which were earlier calculated for elliptic systems with unique states. This made it
possible to prove the existence of unique optimal controls of such conditionally
well-posed elliptic systems with discontinuous states. To remove the constraint
$\int_\Omega ydx = Q$, which is too complex to be allowed for in computational algorithms

of searching for optimal controls, a technique was developed to replace elliptic problems with various perturbations (controls) and such constraint by equivalent boundary-value problems for elliptic equations with interface conditions, where a unique state $y = y(u)$ is defined in the corresponding class of discontinuous functions without this constraint. It is shown that it is possible to set up highly accurate numerical schemes to determine an approximation of the unique optimal control $u \in \cup_{\partial} = \cup$.

The optimal control of the states of the following multicomponent distributed systems was also analyzed:

- Elliptic systems with interface conditions specified on two different disjoint surfaces of discontinuity of spatial bodies (including conditionally well-posed systems)
- Elastic spatial bodies with stresses specified on their boundaries (with a multiple-valued inverse operator of state)
- Thermoelastic state of multicomponent bodies
- Pseudoparabolic multicomponent distributed systems
- Elliptic–parabolic and elliptic–pseudoparabolic multicomponent distributed systems

The optimal control was analyzed for unsteady filtration of a fluid in compressible multicomponent soil and formation of unsteady temperature fields in such bodies with thin inclusions arbitrarily located in space; unsteady filtration of a fluid in compressible multicomponent soil media and formation of unsteady temperature fields in such bodies under conditions of fluid pumping out and thin inclusions with concentrated heat capacity; unsteady filtration of a fluid with dense compressible and incompressible components that reduces to studying the optimal control of states of multicomponent bodies with thin inclusions arbitrarily located in space which are described, respectively, by initial boundary-value problems for parabolic equations with conditions of imperfect contact, for parabolic equations with conditions of concentrated heat capacities, for elliptic–parabolic equations with boundary and initial conditions, and for interface conditions of imperfect contact.

Optimal control of unsteady filtration of a fluid in fractured multicomponent bodies reduces to optimal control problems for systems whose state can be described by initial–boundary-value problems for pseudoparabolic equations with interface conditions of imperfect contact. Various possibilities for optimal control of such systems (internal sources, interface conditions, boundary conditions) under conditions of various observations (of the formation of hydraulic pressure head) over the whole multicomponent body, on its boundary, on a thin inclusion, and of the final state of the system are analyzed. For all the considered problems, the existence of a unique optimal control was proved.

Various oscillatory processes can be described by initial–boundary-value problems for an equation and systems of hyperbolic partial differential equations. The optimal control of systems whose state can be described by initial–boundary-value problems for hyperbolic partial differential equations with interface conditions of imperfect contact was analyzed. Various possibilities of the optimal control of the above systems (uniformly distributed internal sources, interface conditions, boundary conditions) under conditions of various observations of the state of the system (over the whole multicomponent domain, on its boundary, with the final observation of the state of the system and the rate of its variation) were considered.

The optimal control of dynamic processes in multicomponent systems under conditions of viscosity was analyzed for systems whose states can be described by initial–boundary-value problems for pseudoparabolic equations with interface conditions of imperfect contact. Various possibilities of the optimal control of such systems (internal sources, interface conditions, boundary conditions) were analyzed under various observations (of the state over the whole multicomponent body, on its boundary, on a thin inclusion, of the state of the system at the final moment of observations). In all these cases, theorems on the existence of unique optimal controls were proved.

The optimal control of the strain state of an elastic multicomponent body with thin low-strength inclusions (cracks) was analyzed. The optimal control (of bulk forces, stresses on a part of the boundary of a given body) with the observations of displacements over the whole multicomponent body, on a thin inclusion over a jump of tangential displacements, of the normal component of the displacements on a thin low-strength inclusion, and of displacements on a part of the boundary of the body was also considered. Theorems on the existence of a unique optimal control were proved for all the above cases. If the set $\cup_\partial$ of feasible controls coincides with the corresponding complete Hilbert space U, it is possible to use the corresponding classes of discontinuous FEM functions to set up highly accurate computational schemes to find the approximation u_k^N of the optimal control u.

Optimal control of thermostressed state, quasistatic stress–strain state, dynamic stress–strain state of multicomponent bodies and systems described by hyperbolic, pseudohyperbolic equations with concentrated masses, convection–diffusion processes, etc., was analyzed. Optimal control of conditionally well-posed systems of thermostressed state of bodies where thermal processes and the strain state of the body are ambiguously defined was also considered.

For all the above-mentioned multicomponent distributed systems, the existence and uniqueness of the optimal control $u \in \cup_\partial$ are established, and explicit expressions of the description are obtained for the following: a state of the system

$$L(y, u) = 0, \quad y \in H, \quad u \in \cup_\partial, \tag{3.23}$$

the conjugate state

$$L^*(y^*, y) = 0, \tag{3.24}$$

and the inequality

$$(Cy(u) - z_g, Cy(v) - Cy(u))_H + (N u, v - u)_{\cup} \geq 0 \quad \forall v \in \cup_{\partial} \tag{3.25}$$

with the cost functional

$$J(u) = \pi(u, u) - 2L(u) + \|z_g - y(0)\|_H^2,$$

where

$$\pi(u, v) = (y(u) - y(0), y(v) - y(0))_H + (N u, v)_{\cup},$$
$$L(u) = (z_g - y(0), y(v) - y(0))_H,$$

which determine the optimal control $u \in \cup_{\partial}$.

8. Identifying the Parameters of Multicomponent Distributed Systems

As mentioned above, obtaining a reliable information about the development of processes in certain objects by mathematical modeling (including the case of using computer facilities) requires mathematical models to be provided with reliable input data, which is a challenge, including modeling the dynamics of states of naturally multicomponent soil. Therefore, it is currently extremely important to develop efficient numerical methods to identify parameters of various multicomponent distributed systems to solve various problems of rational environmental management, construction, ecology, etc. Developing such means will allow creating various powerful information technologies self-adjusting to objects.

One of the approaches to creating such technologies is to develop gradient methods for the minimization of residual functionals (with the use of known traces of the unknown solutions) based on the theory of the optimal control of states of multicomponent distributed systems [30, 155, 218], with the use of the solutions of direct and conjugate problems at each step of the iterative process.

To implement these methods in the solution of inverse problems, it is necessary to calculate the gradient J'_u of the residual functional

$$J(u) = \frac{1}{2} \sum_{i=1}^{N} \|y(u) - f_i\|_{L_2(\gamma_{iT})}^2$$

at each step of the iterative process on the assumption, for example, that the traces of the unknown solution

$$y = f_i, \quad x \in \gamma_i, \quad i = \overline{1, N}, \quad t \in (0, T)$$

are known on the surfaces γ_i for $t \in (0, T)$. We can find the gradient J'_u from the expression

$$\langle J'_u, \, v - u \rangle \approx (=)(y(u) - z_g, \, y(v) - y(u))_{\mathrm{H}},$$

where the right-hand sides can be constructed with the use of the solutions of the corresponding conjugate problems found earlier in the optimal control theory. Some of such methods of solving parameter identification problems for multicomponent distributed systems are developed, for example, in [155]. This study considers the construction of the gradients of residual functionals for gradient methods of solving parameter identification problems (boundary conditions, interface conditions, thermal conductivity and thermal capacity coefficients, internal sources and their combinations) for parabolic multicomponent systems, complex problems of parameter identification for pseudoparabolic multicomponent systems, complex problems of parameter identification for thermoelasticity problem, etc. The construction is based on expressions (3.23)–(3.25).

3.3   Optimization of Computations and Information Security

The middle of the last century witnessed a tremendous growth of science and engineering, including computational mathematics. The use of computers to solve complex problems arising in all sectors of the economy generated new problems such as solving problems and using computers in the best way, that is, so as to obtain the solutions of certain problems with prescribed characteristics as to the accuracy, speed, and computer memory. All this resulted in a new research area, computation optimization.

The subject of study in computational mathematics is methods to find numerical solutions to various mathematical problems. Together with the traditional subdisciplines such as calculation of values of functions, computational methods of linear algebra, numerical solution of algebraic and transcendental equations, differential and integral equations, and numerical methods for the optimization of functions and functionals, new subdisciplines develop such as theory and numerical methods to solve ill-posed problems, linear and nonlinear programming, theory of data processing systems, etc. Each of these subdisciplines had then its numerical methods. What all these methods and the corresponding computational algorithms had in common is the necessity of their theoretical study to solve the following problems: establish the existence of the solution and the convergence of approximate method, study the rate of convergence, effective estimation of error, computational complexity of approximate method, and information capacity.

These problems were solved for each class of problems, each method in a certain way, and quite often caused severe difficulties which could not always be overcome effectively. It is important to combine these studies and to use the ideas and elements of functional analysis to set up a unified general theory of approximate methods.

In line with the above issues, intensive research started at the Institute of Cybernetics in the 1950–1960s to solve various classes of problems with the application of approximate methods. In particular, the fundamentals of the theory of errors, both of separate ones, which accompany the approximate solution of problems, and of the total error were developed. General schemes of total error estimation were developed and used to estimate the total error in solving classes of problems such as estimation of functions and functionals; statistical processing of experimental data; linear and nonlinear functional and operator equations, which a priori have a unique solution in a given domain; minimization of various classes of functions and functionals; of some classes of ill-posed problems; Cauchy problems for systems of differential equations; etc.

Naturally, the publications of new ideas and results concerning the above problems stimulated the organization of various mathematical forums in order to discuss new results obtained in computational mathematics, to report them to the mathematical community, and to establish new fields of the development of the subject. The management of the Institute of Cybernetics agreed[1] to make the institute a leading organization of such scientific forums.

The first such scientific forums were held in Kyiv in June 1969 (the All-Union Symposium "Accuracy and Efficiency of Computational Algorithms") and in Odessa in September 1969 (the first mathematical school with the same name).[2]

Thus, the period of 1950–1960s can be characterized as the first stage in the scientific development of the theory of computation. The further studies in the theory of computation deepened the knowledge about available numerical methods. For example, for estimation of errors such as irreducible and roundoff ones, the following estimates were obtained: a priori, a posteriori, majorant and asymptotic, and deterministic and stochastic. The generalization and modification of available methods were investigated to optimize the relevant computational algorithms in both accuracy and speed. According to these criteria, order-optimal, asymptotically optimal, and optimal algorithms were defined [22].

Since the solution of all scientific and technical problems reduces to typical problems of computational mathematics (statistical data handling, estimation of functions and functionals, solution of equations, minimizations of functions, and mathematical programming), considerable attention was paid to the quality of the methods of their solution. As an example, we will consider the estimation of the total error for a solution algorithm for a problem of statistical data handling [65].

Methods of probability theory and mathematical statistics are usually applied for data handling. These methods are based on the law of large numbers, according to

---

[1] This was proposed at the Republic Scientific Conference on Computational Mathematics held in spring 1968 at the Uzhgorod State University.

[2] In September 2009, in Katsiveli (Crimea), the V. M. Glushkov Institute of Cybernetics held the 35th International Symposium "Optimization of Computations" devoted to the 40th anniversary of this event.

which, given a considerable number of independent trials, the probability of events is close to their frequency, and the mathematical expectation of random variables is close to the arithmetic mean. However, in practice, one is often restricted to a relatively small number of trials. Hence, there is an additional problem to estimate the accuracy of characteristics obtained from a trial.

Let X_v be the value of a random variable X taken as a result of the vth trial and x_v be the specific value of the random variable X obtained as a result of the vth trial. To determine the total error estimates of expectation m_x^*, variance d_x^*, probability density function of random variables, correlation moments R_{xy}^*, and correlation coefficients r_{xy}^* for random variables X and Y, except for the error estimates of the method, one should additionally analyze the irreducible and roundoff errors. Let us use the estimate $m_x^* = \frac{1}{n} \sum_{v=1}^{n} x_v$ as an example. Suppose that instead of x_v we are dealing with $x_{v,\varepsilon}$ (an approximate value of x_v), and the relationship $D(E_v) = \sigma^2$ is true for the variance of the random variable $E_v = x_v - x_{v,\varepsilon}$. Then the obtain $m_{x,\varepsilon}^*$ $= \frac{1}{n} \sum_{v=1}^{n} x_{v,\varepsilon}$ instead of m_x^*. Assuming that E_v are pairwise independent, according to the Chebyshev inequality, the following irreducible error estimate is valid with probability 0.96: $|m_x^* - m_{x,\varepsilon}^*| \leq \frac{5\sigma}{\sqrt{n}}$; if E_v are pairwise independent and normally distributed random variables with zero expectation and variance $D(E_v) = \sigma^2$,

$$P(|m_x^* - m_{x,\varepsilon}^*| \leq \delta) = \frac{2}{\sqrt{2\pi}} \int\limits_0^{\frac{\delta\sqrt{n}}{\sigma}} e^{-\frac{z^2}{2}} dz = 2\Phi\left(\frac{\delta\sqrt{n}}{\sigma}\right),$$

where

$$m_{x,\varepsilon}^* = \frac{1}{n} \sum_{v=1}^{n} x_{v,\varepsilon},$$

$$\Phi(n) = \frac{1}{\sqrt{2\pi}} \int\limits_0^n e^{-\frac{t^2}{2}} dt,$$

and P is the probability of $|m_x^* - m_{x,\varepsilon}^*| \leq \delta$.

Provided that $(n+1)2^{-\tau} < 0.1$, the estimate

$$|m_{x,\varepsilon}^* - m_{x,\varepsilon,\tau}^*| \leq 1.06 \max_j |x_{v,\varepsilon}| \frac{(n+2)(n-1)}{2n} 2^{-\tau}$$

is true for the roundoff error estimate of m_x^* computed in floating-point arithmetic, where τ is the number of digits in the mantissa. The error may be significant for large n. To avoid this, summing should be executed preferably without rounding off. $X_{v,\varepsilon,\tau}$ are known to be random variables asymptotically (with respect to τ)

uniformly distributed over $\left(-\frac{1}{2}2^{-\tau}, \frac{1}{2}2^{-\tau}\right)$. If $X_{v,\varepsilon,\tau}$ are pairwise independent, the estimate $\left|m_x^* - m_{x,\varepsilon,\tau}^*\right| \leq \frac{5\cdot2^{-\tau}}{2\sqrt{3n}}$ holds with probability 0.96, where $m_{x,\varepsilon,\tau}^* = \frac{1}{n}\sum\limits_{v=1}^{n} x_{v,\varepsilon,\tau}$.

Considering that the law of distribution of $m_{x,\varepsilon,\tau}^*$ is nearly normal, the roundoff error can be estimated more accurately.

In the mid-1970s, M. D. Babich, senior researcher of the institute, studied an important and complex problem of the global solution of nonlinear functional equations by preliminarily separating all the isolated solutions and then refining them by iterative methods. This problem is rather complex and was not solved at that time. The algorithm of separating all isolated solutions (called εs-algorithm (ε-grid)) was developed based on Nemytskii's idea applied to nonlinear operator equations. An algorithm and a program were developed. The results of the numerical experiment using test examples were encouraging. The program found all the isolated solutions to systems of nonlinear algebraic equations in a given bounded domain. The results of the experiment were shown to V. S. Mikhalevich who conducted research in computational mathematics at the institute. He studied the results with great interest and advised to continue the research and extend it to more complex classes of nonlinear functional equations. In this context, it became necessary to develop elements of the general theory of the approximate solution of nonlinear functional and operator equations.

The essence of the developed elements of the general theory is as follows: For an exact (input) operator equation $Tu = 0$, construct a sequence of approximate operator equations $T_n u = 0$ and identify the relationship between them, which implies the proof of the theorem that the solvability of the exact equation yields the solvability of a sequence of approximate equations, beginning with a certain or arbitrary n and the convergence of the method of transition to them, and the proof of the inverse theorem that the existence (for a fixed n) of isolated solutions of the approximate equation yields the existence of the corresponding solutions of the exact equation and a posteriori error estimate. The domain of uniqueness of the solution was proposed to be taken as a closed ball with the center at the approximate solution of the approximation equation and the radius determined from the sufficient conditions of the theorems on the existence of the solution and convergence of the iterative method.

M. D. Babich carried out related research by creating elements of the general theory of approximate solution of nonlinear operator and functional equations [4]. The results of this research were included in his doctoral dissertation, which he defended in 1993. Academician Mikhalevich was a scientific consultant; he was interested in the research and took active part in the discussions throughout the period of the Ph.D. research. Some results of this research were included in joint publications [96].

Let F be a class of problems of computational mathematics and $A(X)$ be a class of computational algorithms used to derive the solution of problems from the class F with the use of input information $I = I(I_0, I_n(f))$, where I_0 is the information about the properties of a problem from the class F and $I_n = I_n(f) = (i_1(f), \ldots, i_n(f))$ is the information about the problem as n functionals calculated on the elements of the problem f.

Let $c(Y)$ be a computer model that incorporates some architectural features of the computer and belongs to a certain class of models $C(Y)$. One of the statements of a computing optimization problem is to find a solution (approximate in the general case) of a problem $f \in F$ under the following conditions:

$$\rho(E(I, X, Y)) \leq \varepsilon, \qquad\qquad (3.26)$$

$$T(\varepsilon, I, X, Y) \leq T_0(\varepsilon), \qquad\qquad (3.27)$$

where $\rho(\cdot)$ is some measure of the error of the approximate solution of the problem $f \in F$; $E(I, X, Y)$ is usually the total error of the approximate solution; X and Y are vector of parameters that characterize, respectively, algorithms and computers from classes A and C; $T(\varepsilon, I, X, Y)$ is the CPU time needed to calculate the approximate solution; ε and $T_{0(\varepsilon)}$ are constraints based on the requirements to mathematical modeling and the properties of input information (volume, accuracy, structure, acquiring method).

An approximate solution for which condition (3.26) is satisfied is called ε-solution; $A(\varepsilon, X, Y)$ is the set of computational algorithms to derive the ε-solution in the given computer model of calculations. A computation algorithm for which conditions (3.26) and (3.27) are satisfied is called T-efficient; $A(\varepsilon, T_0, X, Y)$ is the set of T-efficient computation algorithms in the given computer model of calculations.

In the early 1980s, an important trend in the theory of computation, including computational optimization, was related to complex problems of one- and multicriteria optimization that necessitated theoretical and practical fundamentals of system optimization in both continuous and discrete cases. This led to the development of mathematical methods and software of online optimization, which in turn facilitated fast implementation of research results in the economy, and from the research standpoint, it entailed the creation of the fundamentals of system optimization and its practical implementation.

As a stage of the development of the theory of computation, the 1990s can be characterized as a period of new challenges in ecology and economy that required new approaches to their analysis and solution. It was also the time of rapid development of interactive systems and software to solve the basic classes of problems with prescribed characteristics, where important points were the maximum use of new advances in computer technology (computers of new generation and multiprocessor computers) and the analysis of algorithms and programs by testing them against sets of tasks and sets of input data [5, 96]. All this resulted in systems analysis, a new discipline that combines all the parts of the solution of complex applied problems. In this regard, Mikhalevich proposed to include the division of systems analysis and mathematical programming methods into the program of scientific forums "optimization of computations."

The advent of the new century witnessed a pressing need to solve some scientific and technical problems such as those related to the use of nuclear energy; design of aircraft; modeling and prediction of the economic development of the society;

protection of financial, economic, and industrial information; theoretical and methodological analysis of the reasons of technical and environmental disasters; development of modern technologies of protecting potentially dangerous objects; etc. Timely and effective solution of such problems depends on the further development of the theory of computation, in particular, the development of new efficient (including time- and accuracy-optimal) algorithms, which reduces the computational costs for different models of computation and (or) uncertainty in the numerical modeling. Furthermore, this extends the set and complicates problems that can be solved with limited computational resources.

A new division of information protection methods was introduced in the scientific forum at that time. This was topical because of the increasing number of publications in the field of information security and the use of these results to solve a wide range of formidable problems, which is impossible without involving the reserves of computation optimization. It should be taken into account that computing power increases (the fleet of supercomputers expands; they are combined into networks) and the relevant databases become available. This gives rise to topical issues of information protection by cryptographic and steganographic means, which are based on knowledge-intensive basic research. This applies first of all to creating secure databases, distribution of encryption keys, decryption of information, applying digital signature, and creating cryptographic protocols.

Let us dwell on some results obtained at the departments headed by I. N. Kovalenko, academician of the NAS of Ukraine, and V. K. Zadiraka, corresponding member of the NAS of Ukraine.

The main research areas in the theory of mathematical methods of data security and combinatorial analysis in the department headed by Kovalenko are as follows:

- Analysis of the properties and parameters of random systems of equations over finite algebraic structures: groups, rings, and Galois fields; the development of methods to solve such systems
- Analysis of Boolean matrices, their ranks and determinants; establishing the invariant boundaries for the characteristics of random matrices with respect to the distribution of their elements
- Analysis of logical equations
- Studies in the theory of random arrangements
- Development of special algorithms to decode heavily distorted linear codes
- Studies in the theory of random graphs
- Analysis of the probability characteristics of sequences over finite alphabets
- Development of asymptotic and finite methods to calculate the number of complete mappings of finite sets
- Studies in the theory of finite automata

Kovalenko's studies on the methods of decoding heavily distorted linear codes resulted in new fast algorithms for finding information words. These algorithms are also used in a different interpretation to solve systems of linear equations in Galois fields with disturbing parameters. The algorithm by Kovalenko and Savchuk significantly reduces the amount of computation for decoding and simplifies the solution of systems of linear equations related to the analysis of a certain class of encoders.

The theory of systems of random equations over finite algebraic structures takes an important place in modern discrete mathematics due to its application in the theory of finite automata, coding theory, the theory of random graphs, and cryptography. For example, the solution of this problem becomes of special interest for cryptography, where systems of random equations are probability-theoretic models of combinatorial processes that occur in data protection systems.

The desired probability properties of a linear transformation can be achieved in the simplest way by generating a matrix with independent equiprobable elements. However, to be implemented, such a transformation requires a very complex procedure. Therefore, the question is what is the minimum possible filling of the matrix at which the probabilistic effect is still the same. The answer is given by the probabilistic theory of invariance for systems of random equations over finite algebraic structures.

Reliable operation of computerized data processing systems implies the security, integrity, and authenticity of data that are received, stored, and processed in these systems. Information is usually transmitted through insecure communication channels, such as the Internet. An insecure channel is a channel where nearly everyone can read all the data passing through it, and all the data structures, their formats, and data conversion and processing algorithms are standard and thus well known. In this channel, data can be modified (deleted, changed, new data can be added) or readdressed and information exchange can be imposed under a false name.

The "digital envelope" technology provides the security, integrity, and authenticity of the information by means of symmetric and asymmetric encryption and digital signature.

Symmetric encryption is carried out according to the state standard GOST 28147–89; asymmetric encryption follows the algorithm developed at the V. M. Glushkov Institute of Cybernetics, which passed the state examination at the Department of Special Telecommunication Systems and Information Protection of the Security Service of Ukraine and became a recommended algorithm. A digital signature is calculated according to the national standard of digital signature DSTU 4145–2002, which was developed at the institute and put into effect in 2003.

The technology enables effective information protection in distributed telecommunication systems using insecure communication channels and technologies of electronic documents. In practical terms, this means that the transmission of any messages through insecure communication channels, such as the Internet, unauthorized users cannot read or modify them, even if it has unlimited computing resources and time. The technology uses domestic cryptographic algorithms developed and implemented by specialists of the Institute of Cybernetics and fully meets the requirements of regulatory documents of Ukraine, including the Law of Ukraine "on electronic digital signature."

The department headed by V. K. Zadiraka paid considerable attention to putting theoretical achievements on the general theory of optimal algorithms into practice to improve the performance of two-key cryptography systems and develop stable steganoalgorithms for digital steganography. Regarding the increased performance of two-key cryptography systems, noteworthy is optimization of the performance of the multidigit multiplication algorithm (storing one number requires no less than 1,024 bits). The importance of the problem stems from the widespread use of this operation for data encryption/decryption, electronic signature, and cryptographic protocols. The Karatsuba–Ofman and Schönhage–Strassen algorithms were optimized with respect to the number of operations, which improved by 25% the estimates of complexity. This gain was confirmed by testing the quality of the developed software. In addition, cooperation with the Ministry of Defense of Ukraine produced a hardware–software complex "Arifmetika" that implemented not only fast multiplication algorithms but also algorithms for summation, subtraction, modulo, and modular powers.

Steganography is an information security method that, unlike cryptography, hides the very existence of secret messages. Computer steganographic techniques are widely used to protect information from unauthorized access, combat systems for monitoring and control of network resources, masking software from unauthorized users, authentication, and copyright protection for some types of intellectual property.

The main task of a steganographic system is to place an incoming message in a container in such a way that any outsider could see nothing but its main content.

By steganosecurity is meant a system's ability to conceal the fact of sending a message from a high-skill attacker; ability to resist attempts to destroy, distort, remove, or replace a hidden message; and the ability to validate or invalidate the authenticity of information.

Today, thanks to the popularity of multimedia technologies, digital steganographic methods are actively developed. Multimedia files are usually allowed to be slightly distorted during storage or transmission. The possibility of making minor changes without loss of functionality of the file can be used for hidden messaging. Human senses cannot reliably distinguish between such changes, and special software cannot identify possible minor changes or cannot be built at the current level of steganographic analysis.

To counter passive and active attacks, secure steganoalgorithms were developed. In the case of passive attacks, algorithms based on estimates of rounding error and convolution theorem for discrete functions appeared to be successful. These are so-called spectral steganoalgorithms. In the case of active attacks, secure steganoalgorithms are based on knowledge of the domains of invariance to certain attacks.

The means of copyright protection is digital watermarking (DWM). Existing steganographic systems with DWM are usually used to protect multimedia data such as digital graphics, digital sound, and digital video. However, the DWM technology can be successfully applied to secure information on paper or plastic carriers. This will greatly expand the range of potential users of these technologies. DWM can be applied to any paper documents, confirming their originality, identifying the owner, and monitoring unauthorized copying and copyright.

Finally, consider the task of testing the quality of application software, which emerged at scientific forums on computing and optimization solution initiated by Mikhalevich [5]. After a scientific forum, a number of departments of the Institute of Cybernetics (Mikhalevich, Sergienko, Shor, Ivanov) were assigned a task successfully accomplished in 1989.

It is known that the development of large complex software applications that implement the computer solution of complex mathematical problems or process control, achieving a high level of their operation, is a complicated task. After making programs workable, they should be analyzed to develop necessary or potential quality indices and to find the place of these programs in optimization of computations. The main tasks are to identify the functional capabilities of tested algorithms/programs, areas of differentiable behavior by accuracy criteria, computer time, computer memory, and, on this basis, to execute optimization computation.

The approach to solving the problem of testing the quality of application software is based on the concept of numerical methods for the coordination of theoretical and computational experiments in view of approximation theory, estimates of accompanying errors, and experimental study of computational algorithms/programs closely related to theoretical conclusions and principles. The objects of study are numerical methods and computational algorithms/programs for solving the following classes of problems: statistical data processing, function approximation, solving equations (linear and nonlinear differential, integral, algebraic, and transcendental), minimization of functions, and problems of mathematical programming.

Test results can be used for:

- Quality analysis of algorithms and their evaluation
- Improvement of algorithms with respect to accuracy and speed
- Selection or refinement of parameters of algorithms to guarantee a solution of prescribed quality
- Selection of the most efficient computational algorithms/programs for solving a specific task or class of problems

3.4 Mathematical Modeling and Optimization of Wave Processes in Inhomogeneous Media

Today, an analysis of various wave processes that quickly vary in time and in space is one of the most challenging problems of computational mathematics and mathematical and theoretical physics. The need to develop methods of mathematical modeling of various direct and inverse wave problems is due to the wide range of current scientific and technological problems of geophysics, electrodynamics, fiber optics, ocean acoustics, atmospheric physics, etc. The practical experience in the analysis of problems of this class shows that their effective solution is only possible with the use of modern computer technologies implemented on powerful super-computer complexes. As will be mentioned below, these problems are rather complex and need large amount of computation and online use of very large amount of data.

We will dwell on some formulations of current problems that are typical for this research area and on the methods of their effective solution. Clearly, we will provide some formal problem statements with their explanation. Note also that in what follows, we will pay more attention to the basic methods of solving real problems in this field, whose program implementation formed the software basis of the related computer technology developed under the guidance of V. V. Skopetskii and A. V. Gladkii [20, 64].

It is also important to effectively solve many problems in the protection and sustainable use of oceans. Intensive investigation into the propagation of acoustic waves in two-dimensional and three-dimensional inhomogeneous underwater waveguides received a considerable impetus in the 1980s in connection with the needs of remote sensing and acoustic monitoring of natural environment. Virtually all types of alarm, communication, radar location, and sounding of water body and ocean floor use acoustic waves. In addition to the direct use in the design of modern information-measuring systems, the results of modeling of wave processes allow solving a wide range of interrelated problems such as the formation of prescribed structures of acoustic fields in waveguides, the synthesis of hydroacoustic antennas, etc.

The features of direct and extremum (inverse) problems of sound propagation in inhomogeneous waveguides cause the increasingly wide use of program-algorithmic systems that implement new technology of studying the patterns of wave processes using mathematical modeling. This allows solving, in a broad range of frequencies, a wide class of problems of acoustic energy propagation in real underwater inhomogeneous waveguides at large distances (within hundreds of kilometers) to reflect, inter alia, with allowance for hydrology, absorption factors, bottom relief, and losses on the waveguide wall. Note also that the study of processes of underwater sound propagation in waveguides is closely related to studies in many fields of scientific and technical activity such as oceanography, atmospheric physics, geophysics, marine hydrotechnics, etc.

From the mathematical point of view, the propagation of acoustic waves in infinite underwater waveguides can be described by boundary-value problems for the Helmholtz equation with complex nonself-adjoint operator or initial–boundary-value problems for parabolic equations of Schrödinger type.

In general, for an inhomogeneous medium with the parameters depending on spatial coordinates, the acoustic field of a harmonic point source (the time dependence is assumed in the form $e^{-i\varpi\Psi\psi\rho t}$) can be described by the Helmholtz wave equation with respect to pressure [20, 64]

$$\rho_0(\mathbf{x})\mathrm{div}\left(\frac{1}{\rho_0(\mathbf{x})}\mathrm{grad}\,p\right) + k^2(\mathbf{x})p = -\delta(\mathbf{x} - \mathbf{x}_0), \qquad (3.28)$$

where $k(\mathbf{x}) = \bar{\omega}/c(\mathbf{x})$ is the wave number, $\mathbf{x} = (x, y, z)$, $\rho_0(\mathbf{x})$ is the density of the medium, $c(\mathbf{x})$ is the speed of sound, and $\delta(\mathbf{x} - \mathbf{x}_0)$ is the Dirac delta function concentrated at the point $\mathbf{x}_0$. If we introduce a coordinate-dependent refractive index $n(\mathbf{x}) = c_0/c(\mathbf{x})$, where c_0 is the speed of sound at some point, Eq. 3.28 becomes

$$\rho_0(\mathbf{x})\text{div}\left(\frac{1}{\rho_0(\mathbf{x})}\text{grad}p\right) + k_0^2 n^2(\mathbf{x})p = -\delta(\mathbf{x} - \mathbf{x}_0).$$

To effectively describe the absorption of acoustic energy, the wave number or refractive index is assumed to be a complex quantity with $\text{Im}k^2(\mathbf{x}) \geq 0$ and $\text{Im}n^2(\mathbf{x}) \geq 0$.

In a cylindrical coordinate system (r, φ, z), the acoustic field of a harmonic point source in a medium with constant density can be described by the Helmholtz wave equation with complex-valued nonself-adjoint and variable-sign operator

$$\frac{1}{r}\frac{\partial}{\partial r}\left(r\frac{\partial p}{\partial r}\right) + \frac{1}{r^2}\frac{\partial^2 p}{\partial \varphi^2} + \frac{\partial^2 p}{\partial z^2} + k_0^2(n^2(r, \varphi, z) + iv(r, \varphi, z))p =$$
$$= -\frac{\delta(r)}{2\pi r}\delta(z - z_0), \tag{3.29}$$

where $p(r, \varphi, z)$ is the complex amplitude of sound pressure, $k_0 = \bar{\omega}/c_0$ is the wave number, $v(r, \varphi, z) \geq 0$ is the absorption rate, and $i = \sqrt{-1}$ is imaginary unit.

Wave problems widely use mathematical models of soft, hard, and impedance walls of waveguides. Impedance conditions (boundary conditions of the third kind) also contain complex coefficients and model the loss of acoustic energy. In a medium of heterogeneous density, interface conditions are satisfied, which mean the continuity of the solution and the flow.

To find the unique solution to boundary-value problems for the Helmholtz equation in unbounded domains, it is necessary to use requirements as estimates of the solution or the asymptotic behavior of the solution at infinity. In the general case of space R^n, Sommerfeld radiation conditions

$$p(\mathbf{x}) = O(r^{(1-n)/2}), \quad \frac{\partial p}{\partial r} \mp ikp = o(r^{(1-n)/2}), \quad r = |\mathbf{x}| \to \infty, \quad \mathbf{x} = (x, y, z) \tag{3.30}$$

should be satisfied, where $k = \text{const}$, and the signs "$\mp$" correspond to the time dependence $e^{\mp i\bar{\omega}t}$.

For a wide class of wave processes described by the Helmholtz equation (3.29) in homogeneous and layered inhomogeneous axisymmetric infinite waveguides, the numerical–analytical solutions can be obtained as the sum of normal modes. Using normal modes to calculate the sound field requires an analysis and finding the eigenvalues and corresponding eigenfunctions of some auxiliary self-adjoint (nonself-adjoint in the presence of absorption) Sturm–Liouville spectral problem

$$\frac{d^2\varphi}{dz^2} + k^2(z)\varphi = \xi^2\varphi, \quad 0<z<H, \quad \varphi(0) = 0, \quad \frac{d\varphi(H)}{dz} + \alpha\varphi(H) = 0, \qquad (3.31)$$

where the wave number $k(z)$ and the parameter α are generally complex (Im$\alpha \leq 0$, Im$k(z) \geq 0$).

In the general case (for Im$k > 0$), the eigenvalues are shown to be complex with positive imaginary part, and the corresponding eigenfunctions form a biorthogonal system with the eigenfunctions of the conjugate problem.

If the parameter α in the impedance condition of the spectral problem (3.31) is complex, then the model allows describing the sound field of a point source as discrete mode representations and considering the loss of energy of the water layer in the bottom. The eigenvalues of the spectral problem (3.31) are established to have positive imaginary part if the complex number α has negative imaginary part (Im$\alpha < 0$).

Using the method of normal modes to model sound fields requires the development of highly accurate numerical methods and algorithms to find the eigenvalues and corresponding eigenfunctions of self-adjoint (nonself-adjoint) Sturm–Liouville problem (3.31). Along with traditional difference schemes, of great interest are schemes that make adequate allowance for the strongly oscillating solution of the differential problem. The ideological basis of the developed highly accurate numerical algorithms to solve problem (3.31) is the use of analytical solutions that are locally identical to the solution of the original differential problem and the estimate of eigenvalues for Reξ_n^2 and Imξ_n^2. The qualitative difference of discrete spectral problems from standard network approximations is that they are exact on local analytical solutions and much better approximate the oscillating solution of the differential problem.

Numerical methods, especially the finite-difference method, allow extending the classes of studied acoustic problems in heterogeneous waveguides with variable parameters. Note that directly applying the classical methods of the theory of difference schemes for wave equations with complex nonself-adjoint sign-variable operator in unbounded domains involves certain mathematical difficulties. The features of complex real problems (inhomogeneous and infinite domain of definition, complex-valued solution of differential equations, nonself-adjoint and sign-variable partial differential operator) require new approaches to substantiate the approximation, stability, and convergence of numerical methods.

A topical problem is to develop a technique to design wave processes in two- and three-dimensional domains by preliminary conversion of the Helmholtz differential equation with subsequent use and substantiation of the numerical solution of the Cauchy problem for an approximating elliptic wave equation with complex nonself-adjoint operator. The Cauchy problem for the Helmholtz equation is ill posed in the sense of Hadamard because the solution contains rapidly increasing waves, which leads to numerical error accumulation. Such an approach needs new methods to be developed for the set up and stability analysis of different schemes with complex-valued nonself-adjoint operator.

Mathematical models for the design of low-frequency acoustic field in an azimuth-symmetric medium (for $k_0 r >> 1$) taking into account direct and inverse waves are based on the solution of the Helmholtz equation $p(r, z) = H_0^{(1)}(k_0 r)u(r, z)$ represented in terms of the Hankel function $H_0^{(1)}(\cdot)$, modulated by a sufficiently smooth amplitude $u(r, z)$, which satisfies the elliptic wave equation [209, 210]

$$2ik_0 \frac{\partial u}{\partial r} + \frac{\partial^2 u}{\partial z^2} + \frac{\partial^2 u}{\partial r^2} + k_0^2(n^2(r, z) - 1)u = 0, \quad i = \sqrt{-1}. \tag{3.32}$$

As a result, the problem reduces to the Cauchy problem for Eq. 3.32 with the corresponding boundary conditions. An important result is also the lower (by unity) dimension of the domain since the coordinate along the axis of wave propagation plays the role of "time."

The stability with respect to the initial data is the main issue in the theory of difference schemes [149]. The operator approach was used for the stability analysis with respect to the initial data, and the conditions were established for the stability of explicit and implicit three-layer difference schemes of the solution of the wave equation (3.32) corresponding to the discrete approximation of the ill-posed Cauchy problem in a nonuniform waveguide (a semiband in cylindrical coordinates (r, z)), initial and boundary conditions being taken into account.

The method of parabolic equation is one of the most promising methods in the mathematical modeling of acoustic fields. Its foundations were laid by M. O. Leontovich and V. A. Fok, and it became widely used after the studies by F. Tapert [20, 209, 210, 222, 224]. The method is based on initial–boundary-value problems for parabolic wave equations of Schrödinger type that approximate the Helmholtz equation. This approach is widely used to model the propagation of sound waves over long distances, mostly for unidirectional wave processes.

Parabolic approximations of the Helmholtz wave equation in axisymmetric cylindrical waveguides are based on representing the solution in the form $p(r, z) = = H_0^{(1)}(k_0 r)u(r, z)$, where the amplitude $u(r, z)$ satisfies the pseudo-differential equation

$$\frac{\partial u}{\partial r} + ik_0 u - ik_0 \left(E + (n^2(r, z) - 1 + iv)E + \frac{1}{k_0^2} \frac{\partial^2}{\partial z^2} \right)^{1/2} u = 0, \tag{3.33}$$

which contains square root operator.

A fundamental point in the technology of numerical simulation of unidirectional acoustic processes is the development and correctness analysis of difference schemes for initial–boundary-value problems for parabolic equations with complex-valued nonself-adjoint Schrödinger operator. The technique developed for correctness analysis allowed establishing the stability of proposed difference schemes with respect to initial data, to the right side, and the convergence of the approximate solution to the solution of the differential problem.

Recently, much attention is paid to the analysis of linear and nonlinear problems of active damping (active minimization) of acoustic fields in waveguides [1, 20]. Substantially, such problems are to maximize the power radiated by a discrete antenna system into a far-field zone of the waveguide and to find the coordinates and intensities of point sources of a secondary antenna, whose acoustic field minimizes the field of the primary source in the waveguide. In mathematical sense, these problems reduce to the minimization of a certain performance functional depending on two types of parameters: complex amplitudes of the intensities of point sources and their coordinates. The performance functional specifies the power radiated into the far-field zone of the waveguide or the potential energy of the total acoustic field in some area of the waveguide. The analysis of these problems is complicated by the following circumstances. First, the intensities of unknown point sources are sought for in the field of complex numbers, while the coordinates of sources are scalar quantities. Second, the problems of active minimization are ill-conditioned since, by definition, they are inverse problems of mathematical physics. Finally, these problems are high dimensional for deep waveguides and are problems of mathematical physics that need much computing.

Note that in the case of plane (axisymmetric cylindrical) layer-inhomogeneous waveguides, algorithms of the analysis of extremum acoustic problems make substantial use of highly accurate methods to solve a spectral problem of the form (3.31).

Important classes of application problems related to the formation of sound fields in infinite inhomogeneous waveguides can be solved using the parabolic approximation. First of all, these are problems of the synthesis of acoustic fields that have a prescribed distribution of the field amplitude, intensity, etc.

New mathematical models and integral performance criteria for problems of formation of acoustic fields in two- and three-dimensional infinite inhomogeneous waveguides were proposed. They consist in finding the initial amplitude (phase) such that the sound pressure at some specified subdomain would acquire a prescribed distribution of the amplitude, intensity, etc. In the case of an axisymmetric cylindrical waveguide $G = \{(r, z) : r_0 < r < \infty, \ 0 < z < H\}$, three classes of problems of finding a complex-valued control (amplitude, phase, and amplitude–phase) were analyzed. They are caused by a specific physical meaning and are formulated as the minimization of the integral performance criteria

$$J_1(u) = \int\limits_0^H \{\beta(z)\,|p(R,z) - p_0(z)\,|^2 + \gamma(z)\,|u(z)|^2\}dz,$$

$$J_2(u) = \int\limits_0^H \left\{\beta(z)\left(|p(R,\,z)|^2 - |p_0(z)|^2\right) + \gamma(z)\,|u(z)|^2\right\}dz,$$

$$J_3(u) = \int\limits_0^H \left\{\beta(z)\left(|p(R,\,z)\,|^2 - J_0(z)\right)^2 + \gamma(z)\,|u(z)|^2\right\}dz$$

over the solutions $p(r, z)$ of parabolic wave equations of Schrödinger type.

The technique proposed to analyze a wide class of hydroacoustic optimization problems was tested in solving a number of topical applied problems such as the following problem of focusing acoustic energy on a given area of the waveguide $G = \{(r, z) : r_0 < r < \infty, \ 0 < z < H\}$ with a soft upper and an impedance lower boundaries.

We will present the mathematical statement of the problem of the acoustic field formation in an inhomogeneous waveguide as the minimization of some functional to ensure the prescribed characteristics of the acoustic field, the initial distribution of the acoustic field $u(z)$ being taken as control.

One of the extremum problems is to find $u(z)$ to provide a prescribed distribution of the acoustic field in a certain area of the waveguide, and it reduces to the minimization of the functional

$$J(u) = \int_{\Omega} \rho(z) |p(R, z) - p_0(z)|^2 dz + \int_{\Omega} \gamma(z) |u(z)|^2 dz, \quad \Omega = (0, H), \qquad (3.34)$$

where $p(R, z) = p(R, z, u)$ is the solution of the problem

$$2ik_0 \frac{\partial p}{\partial r} + \frac{\partial}{\partial z} \left(\frac{1}{n(r, z)} \frac{\partial p}{\partial z} \right) + 2k_0^2 (n(r, z) - 1 + iv(r, z) + \mu(r, z)) p = 0, \quad (3.35)$$

$$(r, z) \in G,$$

$$p|_{z=0} = 0, \quad \left(\frac{\partial p}{\partial z} + \alpha p \right) \bigg|_{z=H} = 0, \qquad (3.36)$$

$$p|_{r=r_0} = u(z), \qquad (3.37)$$

which corresponds to the complex-valued control $u(z)$ from a convex closed set $U = \{u(z) \in L_2(\Omega)\}$, and $\rho(z) > 0$ and $\gamma(z) \geq 0$ are given continuous real weight functions. The scalar product and the norm in $L_2(\Omega)$ are defined by the formulas

$$(u, v) = \left(\int_{\Omega} u\bar{v} dz \right)^{1/2}, \quad \|u(z)\|_{L_2(\Omega)} = \left(\int_{\Omega} |u(z)|^2 dz \right)^{1/2},$$

where the dash denotes complex conjugation.

In (3.35)–(3.37), $n(r, z) = c_0/c(r, z)$ is refraction index, $c(r, z)$ is the speed of sound (c_0 is some its value), $k_0 = \varpi/c_0$ is wave number, ϖ is frequency, $v(r, z) \geq 0$ is absorption rate, α is a complex number, $\mathrm{Re}\,\alpha \geq 0$, $\mathrm{Im}\,\alpha \leq 0$, $\mathrm{Re}\,\alpha + |\mathrm{Im}\,\alpha| \neq 0$, and $\mu(r, z)$ is a certain real coefficient.

Thus, the optimization problem is to determine a complex-valued control $u \in U$ such that functional (3.34) attains its lower bound

$$J(w) = \inf_{u \in L_2(\Omega)} J(u),$$

and $p(r, z) = p(r, z, u)$ is a solution of problem (3.35)–(3.37).

Functional (3.34) is shown to be differentiable at an arbitrary point $u(z) \in U$ in a complex space with the scalar product

$$\langle u, v \rangle = \mathrm{Re}(u, v)_{L_2(\Omega)}.$$

With the use of the conjugate problem

$$L\bar{\psi} = 0, \tag{3.38}$$

$$\bar{\psi}\big|_{z=0} = 0, \quad \left(\frac{\partial\bar{\psi}}{\partial z} + \alpha\bar{\psi}\right)\bigg|_{z=H} = 0, \tag{3.39}$$

$$\bar{\psi}\big|_{r=R} = 2\rho(z)(\overline{p(u)} - u), \tag{3.40}$$

in the domain $G_R = \{r_0 < r < R, \ 0 < z < H\}$, where

$$Lv = -2ik_0\frac{\partial v}{\partial r} + \frac{\partial}{\partial r}\left(\frac{1}{n(r, z)}\frac{\partial v}{\partial z}\right) + 2k_0^2(n(r, z) - 1 + iv(r, z) + \mu(r, z))v,$$

real pairs $(\mathrm{Re}\,u, \ \mathrm{Im}\,u)$ are established to be Frechet differentiable in the space $L_2^2(\Omega)$. The gradient of the functional has the form

$$J'(u) = \{\psi_1(r_0, z, u) + 2\gamma(z)u_1, \ \psi_2(r_0, z, u) + 2\gamma(z)u_2\},$$

where $\psi = \psi_1 + i\psi_2$ is a complex-valued solution of the conjugate problem (3.38)–(3.40), $u = u_1(z) + iu_2(z)$.

Thus, to determine the gradient, it is necessary to find the solution of two boundary-value problems for a fixed $u(z)$: First, use the direct problem (3.35)–(3.37) to determine $p(r, z, u)$ and then find the value of the conjugate function.

To solve the extremum problem by gradient methods [9], it is first necessary to calculate the gradient of the functional and to formulate the necessary optimality conditions. In the case of extremum problems without constraints, the optimality condition has the form

$$\mathrm{grad}\, J(w) = 0, \quad J(w) = \inf_{u \in L_2(\Omega)} J(u). \tag{3.41}$$

For the approximate solution of problem (3.41), iterative methods can be used, in particular,

$$\frac{w_{k+1} - w_k}{\tau_{k+1}} + \operatorname{grad} J(w_k) = 0, \quad k = 0, 1, \ldots, \tag{3.42}$$

with the use of different methods to select iterative parameters τ_k. In practice, the iterative parameter in (3.42) is often chosen from the monotonicity condition $J(w_{k+1}) < J(w_k)$.

For constrained optimization problems, the iterative process

$$w_{k+1} = P_U(w_k - \tau_{k+1} \operatorname{grad} J(w_k)) \tag{3.43}$$

is used instead of (3.42), where P_U denotes projection onto the set U.

Algorithm (3.43) can be implemented as follows:

$$\frac{\tilde{w}_{k+1} - w_k}{\tau_{k+1}} + \operatorname{grad} J(w_k) = 0, \tag{3.44}$$

$$w_{k+1} = P_U(\tilde{w}_k), \quad k = 0, 1, \ldots. \tag{3.45}$$

At the first step (3.44), the computation is similar to the unconstrained case; at the second step (3.45), input into the subspace of constraints U is implemented.

The approximate calculation of the gradient involves the numerical solution of the direct boundary-value problem (3.34)–(3.37) and of the conjugate problem (3.38)–(3.40) with complex nonself-adjoint operators. It can be easily seen that the difference schemes proposed to solve the direct problem can be used for the numerical solution of the conjugate problem.

The information technology created for the analysis and optimization of wave processes efficiently implement new mathematical models, methods, and tools for the solution support of real problems of underwater acoustic wave processes. As already noted, the analysis is concerned with the solution of problems of transcomputational complexity, which arise due to extra large waveguides (hundreds and thousands of kilometers), significant amount of data, and the complex structure of the World Ocean environment. The proposed models and methods are practicable; they take into account physical and hydrological features, absorption effects, and boundary relief in a wide range of frequencies.

The new technology also allows the analysis of optimization problems for acoustic processes associated with the formation of acoustic fields in inhomogeneous waveguides of length up to several thousands of kilometers to attain the accuracy of high-order discrete problems (10^8–10^{10}).

The algorithms of the proposed computer technology are parallelized using SCIT-1 and SCIT-2 supercomputers. This contributed to the practical study of wave processes in infinite inhomogeneous waveguides in the most comprehensive physical and mathematical statements.

3.5 Computer Technologies and Analysis of Biological Processes

The great history of genetic studies dates back to the discovery of the laws of inheritance by Mendel in the early 1860s and the rediscovery in the early 1900s. In 2003, on the 50th anniversary of discovering the DNA structure, the Human Genome Project (HGP) was announced to be completed. The genome era has become a reality. DNA is now recognized to be a hereditary material, its structure is understood, and a universal genetic code and efficient automatic genome sequencing methods are developed.

New research strategies and experimental technologies have generated a stable flow of constantly growing complex sets of genome data, the majority of which being available in databases. This has transformed the study of virtually all vital processes depending on the sizes of samples being observed and the number of attributes of objects being studied. Advances in genetics and biochemistry and new computer technologies have provided biologists with highly improved research methods that allow analyzing the functioning of organisms and the state of their health and diseases at the unprecedented molecular level of detailing.

The practical value of new fields of study is obvious. Identifying genes responsible for Mendelian characters formerly required huge efforts of a large team of researchers and often yielded uncertain results. Today, they can be carried out by students in accordance with an established procedure during several weeks, using the access to DNA samples, Internet access to genome databases, and DNA sequencing machines.

The microarray technology has allowed many laboratories to change over from studying the expression of one or two genes per month to the expression of tens of thousands of genes per day. The clinical capabilities regarding the genic symptomatic prognosis for diseases and adverse effects of medications have been rapidly increasing. Genomics opens new prospects in the development of pharmacology toward creating new medications. In the immediate future, the cost of genome sequencing for an individual human will be comparable with that of an ordinary blood test; this will cause radical reorganization in medicine, which will be based on human's genetic passports.

Genetic sequences of approximately 1,000 genomes of higher organisms, microorganisms, bacteria, and viruses have been decoded. As was determined in the Human Genome Project (HGP), the human genome consists of about three billion pairs of nucleotides and contains approximately 25,000 genes. Each DNA chain is a linear sequence of four nucleotides: A (adenine), C (cytosine), G (guanine), and T (thymine). A–T and C–G pairs are known to be complementary pairs of bases that link two strands of a DNA molecule. Genetic information in complementary chains is encoded and decoded in opposite directions. A gene is an individual section of a genome into which the information about one protein is coded. Protein is a linear sequence of amino acids. According to the universal genetic code, each triple of nucleotides defines one of 20 amino acids. In a living cell of an organism, the linear sequence of amino acids becomes spatial very quickly, and this form determines the protein's function in the organism.

Since 2003, the statistical analysis of more than one hundred of genomes (such as that of human being, chimpanzee, mouse, rat, Tetraodon fish, *C. elegans* worm, plants, bacteria, and viruses) has been carried out at the Institute of Cybernetics in order to reveal the patterns governing the encoding of genetic information at the chromosome DNA level [28]. As calculations show, the number of A nucleotides in one DNA chain equals the number of T nucleotides, and the number of C nucleotides equals the number of G nucleotides. Note that the fact that the nucleotides in two DNA chains are complementary does not necessarily mean that the numbers of complementary nucleotides in one chromosome chain are equal.

The relationships for pairs and triples and for longer sequences of nucleotides in one DNA chain have also been revealed. It has been found out that in one DNA chain of a chromosome, the number of occurrences of a certain sequence of nucleotides is equal to the number of occurrences of the complementary sequence encoded in the opposite direction. The resultant relations for the complementary sequences of nucleotides can be written as

$$n(ij...k) \approx n(\bar{k}...\bar{j}\bar{i}), \qquad (3.46)$$

where $i, j, k \in \{A, C, G, T\}$, $\bar{A} = T$, $\bar{C} = G$, $\bar{T} = A$, and $\bar{G} = C$. For example, the number of ACC pairs is equal to the number of GT pairs, the number of TCA triples is equal to the number of TGA triples, and the number of GAGC quadruples is equal to the number of GCTC quadruples. Therefore, the nucleotides encoded in both DNA chains of the chromosome follow the same pattern, and the molecular weights of both chains are equal.

Important repetitions of identical complementary characters have also been revealed in genomes. A computer has calculated the number of individual sequences that consist of identical letters A, T, C, and G. As numerical calculations show, the following relations hold:

$$n(A...A) = n(T...T), n(C...C) = n(G...G).$$

For example, in the chromosome 2 of the human genome, there are three sequences that consist of 50 letters A and the same number of those consisting of 50 letters T, as well as 131 sequences of 10 letters C and the same number of sequences consisting of 10 letters G. Note that the number of different variants of sequences that consist of 20 letters is $4^{20} = 2^{40} \sim 10^{12}$, and for 50 letters, it is $4^{50} = 2^{100} \sim 10^{30}$. That is why the SCIT multiprocessor computer was used for these calculations.

The following important property of two-chain chromosome DNA molecules follows from relation (3.46): The estimates of transition probabilities of Markov chains for both DNA chains calculated in opposite directions coincide.

The amino acid sequence of a protein is formed by translating the four-letter alphabet of nucleotides into the 20-letter alphabet of amino acid residues.

The genetic code is a function that transforms each triple of nucleotides into one of the amino acids. Proteins are synthesized along two DNA chains in opposite directions. Relations such as (3.46) hold also for disjoint triples of nucleotides and sextuples that consist of two disjoint triples. Therefore, one may conclude that amino acid sequences of proteins synthesized along the first chain have the same estimates of transition probabilities as proteins synthesized along the second chain do. Thus, for nucleotide sequences of DNA and amino acid sequences of proteins, recognition procedures that use models of Markov chains can be developed.

Genomes of bacteria have comparatively simple structure: Protein-coding sections are not interrupted with noncoding inserts (introns). This feature of bacteria genomes allows separating out and analyzing separately the amino acid sequences of proteins. The numerical calculations carried out using genomes of bacteria have confirmed the above-mentioned conclusion: The number of amino acids and individual pairs of amino acids in proteins that are synthesized along two DNA chains of bacteria coincide.

The patterns thus obtained substantially supplement modern representations about encoding genetic information in DNA and proteins.

The new international project ENCODE (Encyclopedia of DNA Elements) continues the study of the human genome. The main objectives of the project are to develop an exhaustive and understandable catalogue of all the functional components of the human genome, to determine the biological functions of these components at the level of a cell and organism as a whole, and to explain the mechanism of changes and modifications in genomes. Let us dwell on the above problems, which are also investigated by experts at the Institute of Cybernetics.

1. Exhaustive Identification of Structural and Functional Components Encoded in the Human Genome. Despite the relative simplicity of DNA molecule, the structure of the human genome is extremely complex, and not all of its functions are known. Only 1–2 % of its nucleotides encode proteins. The noncoding part of the genome is subject to active study. It is also assumed to be functionally important and probably contain the main part of regulatory information by supervising the expression of approximately 25,000 protein-coding genes and the operation of many other functional elements such as non-protein-coding genes and a number of key factors of chromosome dynamics. It is much less known about approximately a half of the genome that consists of a multiply repeated sequence and about the other noncoding DNA sequences.

The next stage of the development of genomics is to classify, describe, and explain the full set of functional elements in the human and other genomes. Well-known classes of functional elements such as protein-coding sequences cannot still be precisely predicted based on a nucleotide sequence. Other types of known functional sequences such as genetic regulatory elements are even less clear.

Comparing genome sequences from different forms of living objects has resulted in powerful tools for the identification of functionally important elements. For example, the initial analysis of available sequences of genomes of vertebrates has revealed many protein-coding sequences not known earlier, and the comparison of genomes of a number of mammals has found out a great number of homologous sections in the noncoding domains. Further comparison of genome sequences for many living objects, especially those occupying different evolutionary positions, will significantly improve our understanding of the functions of these sequences. In particular, studying the genome sequences obtained for some forms of animals is decisive in the functional characteristic of the human genome.

Efficient identification and analysis of the functional elements of a genome need computational capabilities to be increased, in particular, creating new approaches to record data sets of increasing complexity and, respectively, a powerful computational infrastructure to save, access to, and analyze such data sets (in essence, the SCIT multiprocessor complex answers these purposes).

2. Studying the Organization of Gene Networks and Protein Regulatory Paths. Genes and gene products do not function independently; they participate in gene networks and molecular systems by ensuring the functioning of cells, tissues, organs, and entire organisms. Determining the structure of these systems, their properties, and interactions is a key factor in finding out how biological systems function. This appears much more difficult than solving any other problem molecular biology, genetics, or genomics ever faced. One of the efficient ways is to study relatively simple model organisms such as bacteria and yeast and then more complex ones such as a mouse and a human.

The information of several levels is necessary to analyze the structures of gene networks and molecular regulatory systems. At the gene level, studying the architecture of regulatory interactions will be necessary to identify different types of cells and to develop methods of simultaneous monitoring of the expression of all genes in a cell. At the level of data on gene products, it is necessary to develop the technologies that will allow real-time measurement of protein's expression, localization, modification, and kinetic activity. It is important to develop and improve the technologies that model gene expression, establish temporary expression of individual proteins, and determine their functions. It is an important next step after determining the functions of all genes and their products.

3. Detailed Study of the Influence of Hereditary Changes in the Human Genome on its Phenotype. Phenotype is the set of all the structural and functional features of an organism. Genetics studies the correlation between the differences in nucleotide sequences of DNA of different organisms and their phenotype features. Significant achievements in human genetics fall within the analysis of variations in phenotype features related to variations in an individual gene. Nevertheless, the majority of the features, including usual diseases and different reactions of an organism to pharmacological means, are more complex and result from the interaction of many genetic and nongenetic factors, that is, of genes and their products and the environment. To explain such phenotype features, changes in the human genome should be described completely, and analytic means that will help to explain the genetic basis of diseases should be developed.

Studying genetic deviations and revealing how they influence the functioning of specific proteins and protein compounds will bring new understanding of physiological processes in normal and disease state.

4. Explaining Evolutionary Intraspecific Changes and their Mechanisms. A genome is a dynamic structure modification prone to modification during evolution. Genome changes that have resulted in the development of a human being are only a small flash in the large window of evolution throughout of which the modern biosphere was formed during billion years of trials and errors. To completely understand the genome functions, one should distinguish between individuals of one species and study a number of fundamental processes.

Interspecific comparison of genome sequences is important to identify functional elements in genomes. Determining intraspecific differences in these sequences will help to find the genetic basis of phenotype features and to reveal the patterns of mutation processes. Mutations lead to long-term evolutionary changes and may cause hereditary diseases. Nowadays, our understanding of DNA mutations and reparations, including an important role of environmental factors, is not complete.

To successfully solve problems in molecular biology and genetics, a new mathematical apparatus should be developed. Using a finite set of known analytic functions, it is impossible to construct exact mathematical models that describe the functioning of an organism at the cell–molecular or gene level.

Using a finite set of parameters (or variables), it is impossible to predict the exact state of complex biological systems since there always exist many latent parameters unobservable because of certain technological constraints. Moreover, processes in animate nature are influenced by various random factors. That is why DNA and proteins are described by random nucleotide and amino acid sequences.

The apparatus of probabilistic models is recognized to be most adequate in describing processes in wildlife. For example, according to the leading scientific journal, *Nature Review Genetics* 5, 2004 (The Bayesian Revolution in Genetics), applying the Bayesian approach has caused revolutionary changes in the studies in biology and genetics.

Scientists from the Institute of Cybernetics carried out efficiency analysis of the Bayesian approach to the solution of complex discrete recognition and prediction problems [28].

The structure of the object description turned out to be the key point in solving recognition problems. If it is known, Bayesian classification procedures can easily be constructed, as is the case for Markov chains, Bayesian networks, and independent attributes. A recognition problem is considered from the point of view of minimizing the average risk:

$$I(A) = \sum_{x \in X} \sum_{y=0}^{1} (A(x) - y)^2 P(x, y), \qquad (3.47)$$

where $A(x)$ is a recognition function, y is a state of the object, and $P(x, y)$ is unknown probability distribution. For example, for Boolean vectors $x = (x_1, x_2, ..., x_n)$, the number of different recognition functions $A(x)$ is 2^{2^n}, that is, the problem of minimizing the average risk (3.47), is of superlarge dimension. For $n = 10$, this number exceeds 10^{300}; therefore, it is impossible to approximate such a set of functions by a finite set of known continuous functions.

The recognition problem in formulation (3.47) is a generalization of classical problems, which are solved based on the least-squares method, and not one but several states of the object can correspond to one observation.

Note that minimizing the average risk (3.47) is more complex than any NP-complete problem since a known recognition function cannot easily be tested (in a polynomial number of operations in input parameters of the problem), and averaging over all 2^n vectors x should be carried out in formula (3.47). At the same time, Bayesian classification procedures are shown to efficiently (optimally) solve this challenge for individual structures of object descriptions.

The direct approach to constructing efficient recognition methods is chosen. Instead of rather complex minimization procedures for empirical risk or other performance functionals, a simple Bayesian classification rule is used to solve high-dimensional problems. Error estimating is a nontrivial problem in the discrete case since it is not known beforehand how scopes of classes, attributes, and values of attributes will appear in the error estimates of the Bayesian classification procedure. The input parameters are shown to linearly appear in the error estimates of the procedure. The lower-bound estimate of the complexity of the class of problems coincides with the upper-bound error estimate up to an absolute constant, that is, the Bayesian classification procedure is optimal. Thus, the problem of minimizing the average risk for discrete objects with independent attributes can efficiently be solved using Bayesian classification procedures.

Efficient recognition procedures have been constructed for an important class of objects described by Markov chains. Error estimates for Bayesian classification procedures based on Markov chains are obtained. They are similar to the estimates for independent attributes. These results follow from the analysis of the properties of estimates of transition probabilities.

The theory developed underlies the solution of topical applied problems such as the challenge of predicting protein spatial structure. As numerical calculations for 25,000 proteins have shown (with the information taken from the Worldwide Protein Data Bank), the application of the developed Bayesian classification procedures using Markov chains provides the 85% average accuracy of protein secondary structure prediction, which exceeds the accuracy of modern methods available. The calculations were performed using the SCIT cluster computer. Bayesian procedures were effectively applied to predict protein secondary structure by mainly using the conclusions on the equality of transition probabilities of Markov chains for complementary chains of chromosome DNA and on the equality of transition probabilities of Markov chains for proteins synthesized from two complementary DNA chains.

Recently, new efficient procedures have been developed worldwide to predict protein spatial structure. A review of available approaches shows that the most promising are the approaches based on probabilistic models (models of Markov chains, Bayesian networks, conditional random fields, etc.).

Bayesian classification procedures were used at the Institute of Cybernetics to construct a diagnostic system for cerebral glioma progression analysis. Efficient software has been developed to process the information on the data analysis of erythrocyte sedimentation rate. The information was provided by the A. P. Romodanov Institute of Neurosurgery of the AMS of Ukraine.

Probabilistic dependence models (directed acyclic graphs, Bayesian networks) were analyzed at the department of inductive modeling and control methods directed by A. M. Gupal. The advantages of such models are their clearness, compactness, the capability to represent cause–effect relations, and the computational efficiency of probabilistic inference procedures. These properties provide the efficient solution of problems in medical, engineering, and computer diagnostics, language recognition, and data analysis in econometrics and bioinformatics. The problem of inference of the structure of a Bayesian network from data is known to be NP-hard. Thus, the well-known methods face severe difficulties when there are several tens of system variables.

The identification of the structure of a Bayesian network is based on searching for separators (the facts of conditional independence among variables of the model). Since complex enumeration problems arise in this case and cumbersome statistical criteria are used for the independence recognition, the problem of obtaining minimum separators arises. This is also important in problems of inference of consequences of observations in graphic models since information propagates in the structure of the model through separators. Efficient approaches to constructing minimum separators have been found. The results open up new capabilities to simplify the solution of the above problems and to construct efficient methods to derive the structure of Bayesian networks from the known statistical data and the obtained set of minimum separators.

Nowadays, Bayesian networks are used in the world to study gene networks. In cells, genes interact with each other, and such a system of interacting genes can be described as a directed graph. A gene network is a graphic representation of information interactions among genes, and constructing the structure of gene networks is one of the major problems in bioinformatics and systems biology. The structure of gene networks is not known a priori and is estimated based on the data obtained with the use of modern technologies of DNA microarray gene expression data as a result of observations of gene interaction (influence of one gene on the expression of others).

There are solid grounds for believing that it is the development of the Bayesian approach that will play an important role in unveiling important mechanisms of the wildlife, biology, and genetics.

3.6 Economic Problems of Transcomputational Complexity

Processes that have occurred in the economics of both developed and post-socialist countries beginning with the 1970s are characterized by increasing complexity, strengthened contradictions, growing influence of various risk factors and uncertainty, as well as by opposite tendencies of the globalization of economy on the one hand and new national models of economic development on the other hand. Nowadays, the fields that are not directly related to the production of goods (science, education, and financial services) play an increasing role; at the same time, we have examples of successful development of countries that stake on "traditional" industrial production and even on the extraction of raw material. Complex and interrelated economic phenomena need highly efficient managerial decisions made under modern conditions. Such decisions are often made in essentially new situations, where a decision maker cannot use his previous experience and act by analogy. At the same time, when making such decisions, one should take into account numerous factors, both structured and hardly formalizable. This is virtually impossible without applying mathematical methods such as economic and mathematical modeling and other components of modern information technologies.

Applying mathematical methods to solve economic problems, as well as problems from other fields of knowledge, often leads to calculations of transcomputational complexity. Some of such problems (production transportation and optimal design problems) were discussed in the previous chapters. In the present section, we will dwell on several problems that are related to urgent economic problems and allow applying modern multiprocessor techniques and parallel computing.

The support of budgetary process is one of the major problems of governmental bodies at all levels. The further economic development of the country strongly depends on the efficiency of its solution. Features of transition economy such as the considerable share of end users that obtain incomes from the state budget, the great volume of indirect state grants and productive state investments into manufacture, the instability of state fiscal system, etc., make the macroeconomic forecast (based on which the volume and structure of the budget revenue are determined) strongly dependent on the volume and structure of budget costs, which are planned depending on the revenue. Thus, macroeconomic forecast and budget planning problems should be solved simultaneously. It is this approach that underlies the budgeting-supporting information technologies started to be developed at the Institute of Cybernetics in the early 1990s by the initiative of Academician Mikhalevich. This resulted in an integrated information-analytic system to be used by executive authorities involved in budgeting (primarily, the Ministry of Finance) and legislative bodies that update the budget and supervise its execution, in particular, the Budgetary Committee of the Verkhovna Rada. Since this system should commute with the existing computer systems of the above-mentioned governmental structures, the principles of safety and high reliability should be observed; the system is based on a unified approach and mainly on domestic studies and technologies.

The scientific results obtained in the last decade at the Institute of Cybernetics, such as macroeconomic models that take into account specific features of the economy of Ukraine and relevant modern computation methods, underlie this system. The basic novelty of the proposed development is that the actions of several pricing mechanisms inherent in different forms of management are considered simultaneously, and balance and econometric dynamic macromodels are integrated within the framework of a certain complex of models. This allows applying a systems approach to the analysis of processes that occur in production and finance during market transformation. All these models lean upon a common information background. To perform variance calculations, use is made of scenarios that are alternative information sets that differ in the assumptions on the values of certain input parameters of models and economic specifications such as taxation rates. Each of such sets is sufficient to carry out calculations over the whole complex of models. The scenarios are stored in a special database structured according to the requirements for the search of necessary information sets according to certain attributes such as assumption on the action of various pricing mechanisms, the use of certain forms, and specifications of taxation, which allows applying methods of scenario analysis and prediction.

Using a complex of models considerably increases the amount of computation as compared with the computations carried out for individual models. The large dimension together with hierarchical relations among variables is integrally inherent in the above-mentioned models. The state budget implies a great number of expense items partitioned quarterly and monthly. The components of the consolidated budget are also 26 regional budgets, hundreds of budgets of regions and towns of regional subordination, thousands of budgets of local self-government institutions, and budgets of special (retirement, disablement, etc.) funds. Each of these budgets is also characterized by numerous expense items. One should also consider a great number of scenarios of further course of events in unstable financial and economic situations. Thus, calculations using a set of models of information-analytic budgeting-support system involve finding the values of hundreds of thousands of variables. At the same time, there is limited time to carry out such calculations, especially if the task is to assess the consequences of verbal proposals during the discussion of a draft budget. Thus, a typical *problem of transcomputational complexity* arises. When solving it, one can use the above-mentioned hierarchical relations among variables, which allow parallel computing on the chosen complex of models and applying multiprocessor technique such as cluster supercomputers, developed over recent years at the Institute of Cybernetics, and advanced telecommunication means as the technical basis in creating the proposed system.

Such a system allows the following:

- To predict the execution of the major aggregated indices of the state budget and its components for various alternatives of the macroeconomic policy, in particular, to estimate monthly the revenue of the state budget for the major items of income

- To predict the consequences of changes in the structure of expenses of the state budget in order to operatively analyze the proposals arriving during its completion in Verkhovna Rada
- To study the mutual influence of the structure of the state budget and the main specifications of taxation and tariff–customs policy on the one hand and dynamics of macroeconomic indices on the other hand
- To assess the influence of the forms of taxation, the volume and structure of expenses of the state budget on the solvent demand of various groups of consumers, and the realization of various types of products, export, and import
- To predict the branch-detailed dynamics of prices, profit, and expenses of manufacturers and consumers (in the large and for aggregated groups) for the terms of validity of the law of Ukraine on the state budget for the current year
- To analyze short-term consequences of the application of different approaches to price control, including those by direct or indirect (e.g., through tax concessions) subsidizing of manufacturers and consumers
- To estimate the influence of variations in hryvnia exchange rate on the export, import, and domestic production of the major types of production
- To predict the variations in the supply of money and other monetary macroparameters under certain budgetary policy
- To analyze mutual settlements among the state budget, local budgets, and special budgetary funds and to predict the execution of these components against aggregates
- To predict the volumes and branch structure of nonpayments for certain price dynamics
- To estimate short- and intermediate-term macroeconomic consequences of the application of various strategies of nonmonetary budget deficit cover and service of public debt

The models and information technologies that are a part of the system lean upon the modern developments from the economic theory (such as equilibrium analysis) and on the analysis methods for systems of discrete, nondifferentiable, and multicriteria optimization, simulation, and scenario prediction and the analysis of fuzzy sets, which are actively developed at the Institute of Cybernetics.

Along with modeling problems as such, the support of budgeting is also accompanied by auxiliary problems, some of them being of transcomputational complexity. This is, for example, the problem of compressing superlarge data arrays to be then transmitted through telecommunication channels with limited throughput. Such a problem arises, for example, in coordinating the components of a consolidated budget with regional representative bodies. The compression implies replacing an array of economic data with analytic expressions (approximants) with a small number of parameter coefficients. Only the values of these parameters are transmitted through the communication channels.

To derive such approximants, algorithms based on the Remez method of successive Chebyshev interpolations were proposed. These algorithms allow a highly accurate solution of both the problem of obtaining the approximant ("direct" approximation problem) and the problem of estimating the values of the initial array ("inverse" approximation problem).

Note that implementing the above-mentioned algorithms to compress large arrays of irregular data (it is such information that is mostly transmitted in budgeting support problems) requires so significant amount of calculation that it generates problems of transcomputational complexity. Such problems can also be solved by paralleling using multiprocessor complexes. For example, algorithms of the best Chebyshev approximation are used to compress large one-dimensional vector arrays (up to 10,000,000 possible values) in order to obtain a small number of parameters of approximating expressions to be applied in problems of predicting macroeconomic parameters during the preparation of the budget of Ukraine.

Problems of transcomputational complexity related to language recognition also arise in the subsystem of computer-aided stenographing of sessions of the Budgetary Committee of Verkhovna Rada developed as a part of the information-analytic budgeting support system. By the way, algorithms used for the recognition are similar to the methods of sequential analysis of variants discussed in the first chapter. Modern high-efficiency computer facilities allow solving language recognition problems automatically (with participation of stenographers) in reasonable time.

The range of economic problems of transcomputational complexity being solved at the Institute of Cybernetics is not restricted to budgeting support problems. To analyze the changes that occur in productive forces of the society, models of input–output balance with variable direct cost coefficients can be applied. Academician Glushkov was one of the first to study these models. He tried to apply them to determine the technological changes that would allow reaching a prescribed level of ultimate consumption under limited resources. Actually, the question was searching for the ways of intensive development of the economy. The calculations involving these models with the use of a detailed commodity nomenclature required high-dimensional inverse matrices to be found. Thus, problems of transcomputational complexity were dealt with too. Applying the Jordan–Gauss method of successive changes of the matrix of direct costs that would concern only one column at each step, Glushkov proposed an original algorithm to solve this problem [21].

The further development of this class of models in view of the specific features of transition economy results in optimization statements. To determine the ways of economy intensification, one can use P. Straff's [200] definition of the Keynesian multiplier m (one of the major macroeconomic indices, which reflect the response of production output to variations in the final demand) according to the following formula:

$$m = \frac{1}{1 - kq(E - A)^{-1}\alpha},$$

where A is the matrix of direct cost coefficients, α is a vector representing the structure of individual consumption and domestic investments, q is a vector determining the share of wages and other incomes of ultimate consumers in the price of regional production, and k, $0<k<1$, is a constant with the values being sufficiently small for $m>0$ to hold.

The problem of choosing the variations in the matrix A that would maximize the multiplier under certain resource constraints is a high-dimensional nonlinear (nonconvex) programming problem. The number of variables in this problem is proportional to the squared number of branches taken into account in the input–output balance. For balance sheets drawn up in Ukraine, this number does not exceed 38; nevertheless, it is obviously small to analyze technological changes. For example, in the available balance sheets, the production of radio electronics and heavy engineering industry are united (along with some other so much different productions) into one branch. For sufficiently reliable results, the number of branches should be at least about 150; the number of variables (components of the matrix A) will be of order of tens of thousands in this case. Thus, an optimization problem of transcomputational complexity arises in this case too.

Note that finding points of global extremum in high-dimensional nonsmooth nonconvex optimization problems such as the above-mentioned optimization problem is a challenge and is not completely solved yet. Therefore, when choosing the numerical algorithm, it is expedient to consider the search for a point that would satisfy the necessary conditions of local extremum. To find such a point, an efficient method for the optimization of nonsmooth functions is necessary. The r-algorithm considered in the second chapter satisfies this requirement.

The results of the solution of such a problem are advisory. Moreover, it contains a number of ambiguous parameters. Therefore, it is of interest to carry out calculations in a dialogue mode where the user can change some model parameters and observe the set of local–optimal solutions that will be obtained in this case. Such an approach allows estimating the dependence of solutions on varying parameters. Nevertheless, a dialogue mode is computationally and resource intensive, the computing scheme being the same for all the initial approximations and the values of the varying parameters. It allows paralleling the computing process according to the input information, which is one of the most simple and convenient paralleling. Thus paralleled procedures can be implemented, for example, using modern multiprocessor systems such as SCIT clusters created at the Institute of Cybernetics.

Chapter 4
Problems of Modeling and Analysis of Processes in Economic Cybernetics

Abstract Important aspects of investigating complicated problems of transition economy are considered, which are especially topical at the present time, in particular, problems of development of information technologies for the determination of ways of energy conservation in the economy of Ukraine. In this case, attention is given to the investigation of the interactions between the real and financial sectors of the country and also to the consideration of features of transition economy that include the abundance of different forms of imperfect competition (monopolism, oligopoly, monopsony, etc.), simultaneous action of several mechanisms of price formation, and various large-scale manifestations of "shadow" economy. In constructing models of such problems, problems of transcomputational complexity often arise, and modern supercomputers and methods of parallelization of computations and optimization are widely used to solve them.

4.1 Modeling of Transition Economy

Summarizing the experience of solving specific economic problems, which were discussed earlier, Mikhalevich paid great attention to the development of *economic cybernetics* – a discipline that deals with the development and application of information technologies to manage economic entities and processes. Doing research in this field, in some of his works Mikhalevich developed the modern notion of informatics and its interaction with related disciplines such as computer science, systems analysis, mathematics, economics, etc. He studied economic cybernetics as a complex scientific discipline covering all aspects of the development, design, creation, and operation of complex economic systems and their application in various fields of social practice, ways of the formalization of information for computer processing, and development of information culture of the society [97, 107]. The recent decades are characterized by rapidly developing methods of generating highly efficient information processes and systems, establishing the fundamentals for applying intelligent information technologies and systems analysis methods in different subject areas such as

I.V. Sergienko, *Methods of Optimization and Systems Analysis for Problems of Transcomputational Complexity*, Springer Optimization and Its Applications 72, DOI 10.1007/978-1-4614-4211-0_4, © Springer Science+Business Media New York 2012

macroeconomic modeling, management of state institutions, national economic complexes, etc. Research teams that developed the theoretical and methodological basis of informatization and used it to elaborate complex state informatization programs began to appear in the early 1980s. In 1993, Mikhalevich initiated the elaboration of The Concept of the State Informatization Policy and Guidelines of the Informatization of Ukraine. These documents raised the problems of reorienting informatization to market economic reforms. During the preparation of these documents, his concept of the necessity of state approach to solving the problem of informatization of Ukraine received a wide response, and specific measures were taken to develop appropriate governmental regulations [127]. The basic idea of the informatization of Ukraine formulated in these documents meets the main strategic objectives of the state: to accelerate reforms and overcome the crisis in the economy and other branches. Mikhalevich thought that in this connection, the society had to learn modern information culture, means and tools of informatics, and systems analysis as soon as possible.

In his last years, Mikhalevich supervised research activities on the application of economic and mathematical modeling, systems analysis, computing, and information technologies for studying the processes of transformation of command economy into market one.

The situation in the mid-1980s in the former USSR required profound socioeconomic reforms, much needed especially by experts in cybernetics and computer engineering. The developed countries rapidly increased their lead in this field; the majority of domestic scientific, technical, and technological developments at the branch level duplicated not the best and newest systems (recall the ES series of computers). Brilliant theoretical results obtained by academic science were not implemented in the industry. At that time, the system of economic relations hampered the development of informatics. Officials at all levels tried to hide the true situation from their leadership and turn it to their advantage. Quite illustrative was the history of automated control systems, which was forcefully sabotaged at all levels. Experts in cybernetics and information technologies were one of the first to understand that the system of economic relations required immediate radical changes and that further scientific progress was impossible without initiative and interest of business entities. It was not fortuitous that the policy of comprehensive changes in the economy and society proclaimed by M. S. Gorbachev was enthusiastically acclaimed by both leading scientists and the whole country. At the same time, already the first steps of reforms showed casual attitude to planning reforms. There was a false impression (perhaps forced) that there were no problems in market economy, the concept of self-regulation was replaced with uncontrollability, and the necessity of detailed elaboration and justification of economic policy in transition period was denied.

Undoubtedly, certain political and social groups interested in this and tried to take advantage of the chaos for self-enrichment and seizure of property. However, as subsequent events showed, most of them came off a loser. Haphazard implementation of reforms deepened the system-wide crisis and exacerbated contradictions, and generated new problems. Under these conditions, a number of leading scientists of the former USSR initiated the study of the transition from planned to market economy.

The director of the Institute of Cybernetics of the Academy of Sciences (AS) of the UkrSSR and chairman of the Systems Analysis Committee of the USSR, Academician Mikhalevich could not stay aside. Being an excellent expert in system research, he realized that the study of the economic system that emerged due to the reforms and was called "transition economy" had to widely apply systems analysis, mathematical modeling, and modern information technologies. With his active participation, the Economic Transition and Integration (ETI) research project was initiated at the International Institute for Applied Systems Analysis (IIASA) in the mid-1980s. As already mentioned, this institute is the center of coordination of fundamental research in global problems, a structure that combines the efforts of various scientific schools and institutions. This project gave an impetus to research that could predict the further course of events and formulate recommendations for management decisions. Unfortunately, this did not happen. But the project was a "campaign against market ignorance" for many economists, in both theory and practice, from Russia and Central European countries. Later on, some of them made a significant contribution to the reforms. However, the actual research conducted under the project did not always answer the questions that arose during the transformation of planned economy.

The very first attempts to introduce market mechanisms into the economy of postsocialist countries intensified the deregulation processes and clearly demonstrated the importance of consistent and theoretically substantiated economic, social, and information policy. It had to be based on basic research and quickly respond to changes in society. At scientific and practical conferences and round-table meetings in the late 1980s, Mikhalevich repeatedly emphasized that this can be achieved by complete computerization of the society.

As already mentioned, in 1993, Mikhalevich initiated a discussion within the scientific community of pressing problems of the development of information society. During this discussion, he emphasized that the task of informatics is not restricted to providing necessary information for decision making. Modern computing and information technology should help to provide comprehensive implications of each of the existing alternatives, propose alternatives for effective actions in the current situation, and thus to actively support decision makers. It was his point of view on the problem of reorienting the informatization to market economic reforms, this view was embodied in a number of projects meant to give a new impetus to the development of the national cybernetics, informatics, and approaches to studying the information needs of society and to creating organizational mechanisms for elaborating and implementing the state scientific and technical informatization policy. These projects were implemented in the following areas:

1. Development of the mathematical models of processes that occur in transition economy
2. Development of information technologies for decision support
3. Creation of software and hardware systems that implement the obtained theoretical results

Monetary control was an urgent and important task under the conditions of intensifying crisis in the financial sector in the early 1990s. Inflation and increasing nonpayments were the most dangerous forms of the system-wide financial crisis; therefore, investigations in the first of these areas began with the modeling of inflation under the specific conditions of transition economy. In particular, the following macroeconomic model was considered and analyzed in [125].

Let $x(t)$ be the real gross national product at time t (the model assumes that time varies continuously on an interval $[0, T]$). Denote by a the share of intermediate consumption in the total social product, by $y(t)$ the value of real GDP at time t, and by W the share of gross investment in GDP. Let W_0 be the share of the investment that provides simple reproduction of worn and torn out means of production (depreciation). Suppose that the growth rate of the gross national product is proportional to the net (of depreciation) investments, b is a coefficient of proportionality. Then the growth of the gross national product is described by the equation

$$\frac{dx}{dt} = (1 - a)b(W - W_0)x. \tag{4.1}$$

Denote by $R(t)$ the share of the GDP spent for final consumption at time t.

According to the assumptions, $R(t) = (1 - W)(1 - a)x(t)$; considering Eq. 4.1, we get the ratio

$$\frac{dR}{dt} = b_1(W - W_0)R, \tag{4.2}$$

where $b_1 = (1 - a)b$.

Denote by $S(t)$ the demand at the time t and assume that the price index $p(t)$ varies proportionally to the difference between the supply and demand (i.e., the consumption share of the GDP):

$$\frac{dp}{dt} = m(S - R), \tag{4.3}$$

where m is a predetermined factor of proportionality.

The demand is determined by total income and savings of consumers $D(t)$. Assuming that the incomes are proportional (with the coefficient q) to the nominal GDP (i.e., $p(t)y(t)$), we obtain the following equation for $D(t)$:

$$\frac{dD}{dt} = qpy - p \min(S, R).$$

This equation and the ratio $S = D/p$ yield

$$\frac{dS}{dt} = \frac{qR}{1 - W} - \min(S, R) - m\frac{S(S - R)}{p}. \tag{4.4}$$

The model consists of differential Eqs. 4.2, 4.3, and 4.4 supplemented with the initial conditions

$$S(0) = S^0, \quad R(0) = R^0, \quad p(0) = p^0. \tag{4.5}$$

System (4.2), (4.3), (4.4), and (4.5) is nonlinear, with a nonsmooth right-hand side of one of the equations. Thus, it cannot be solved by analytic methods, its numerical solution is also not always reasonable because of the high level of aggregation and uncertainty of input data. Under these conditions, the results of modeling should provide qualitative conclusions on the processes under study. In this connection, studying model (4.2), (4.3), (4.4), and (4.5) was focused on its qualitative analysis, which allowed making the following conclusions [125].

1. The system being analyzed may be in equilibrium under constant prices only for certain ratios among the parameters q, W, W_0, and b. Since these parameters are governed by different economic mechanisms in case of transition economy, it is difficult to coordinate their values.
2. Also, market pricing mechanisms described by Eq. 4.3 do not always ensure equilibrium. The analysis of the phase portrait of the system (more precisely, Eqs. 4.3 and 4.4) showed that for $q/(1 - W) - 1 < q/(1 - W) < (W - W_0)b_1$, supply permanently exceeds demand, and if $(W - W_0)b_1 < q/(1 - W) - 1 < q/(1 - W)$, demand exceeds supply. In the former case, there is a crisis of over-production, in the latter case, there are permanent inflation processes, which are called "inflationary crisis of demand." Only if $q/(1 - W) - 1 < (W - W_0) b_1 < q/(1 - W)$, market stability can be achieved due to market pricing mechanisms.

Note that national and foreign researchers repeatedly used model (4.2), (4.3), (4.4), and (4.5) and its modifications; for example, it was used in [2] to analyze the consequences of innovative changes in the economy. Later, the model was modified to allow for exogenous variation in prices because of inflation of costs and the impact of variations in money supply on prices [111]. In this model, prices are defined by the following relation rather than Eq. 4.3:

$$p = \max(p_1, \hat{p}, p_2), \tag{4.6}$$

where the price component p_1 is determined by the variation of supply and demand and should satisfy an equation similar to Eq. 4.3:

$$\frac{dp_1}{dt} = m(S - R); \tag{4.7}$$

$\hat{p}$ is the component determined by the processes of cost inflation (and is assumed to be a known function of time t); p_2 is determined by variations in money supply $M(t)$ and should satisfy the modified Cagan equation

$$\ln \left(\frac{M}{p_2}\right) = \alpha \frac{dP}{dt} p^{-1} + \gamma, \tag{4.8}$$

where $\alpha > 0$ and γ are some predefined constants.

The model (4.2), (4.4), (4.5), (4.6), (4.7), and (4.8) was used to analyze the inflation processes in Ukraine in 1990 through 1994. The calculations based on the model showed that the price increase because of cost inflation was the determining factor in these processes, although the impact of monetary emission was also significant sometimes. Inflationary crisis of demand was characteristic of only the initial stages of inflation (until autumn 1992).

To analyze the mechanisms of cost inflation, in the early 1990s, Mikhalevich initiated the formation and study of fundamentally new balance models, where the input–output equations in stable (relative) prices were supplemented with equations for current prices and with income–expense balance ratios for separate groups of producers and consumers. This approach allowed for the peculiarities of the transition: market and nonmarket methods of product distribution, simultaneous effect of different pricing mechanisms, and different forms of ownership. Since fast transition processes were studied, the dynamic models were considered as systems of nonlinear differential equations that cannot usually be solved by analytic methods. However, combining methods of qualitative analysis, decomposition, some analytic transforms, and numerical experiments made it possible to draw conclusions on the properties of solutions of such systems and to compare them with actual processes that occur in a transition economy.

The above model was used to analyze the inflationary scenario, which was named "structural inflationary crisis." It developed since there were industries with high production costs and high demand for their products from other industries. An analysis of the 1990s intersectoral balance made it possible to identify such industries. These are primarily fuel and energy complex and transport. Since the effect of these industries on cost inflation and related payment crisis was the strongest, it was decided to start structural and technological transformations in the economy of Ukraine in exactly these industries.

According to the second field of research, related to the development and application of computer and information technologies, hierarchy analysis methods were modified and applied in comparing alternatives of social and economic development. These methods were substantiated based on fuzzy set theory and multicriteria optimization algorithms, which helped to explain some paradoxes that occurred in solving applied problems.

Since the mid-1990s, the Institute of Cybernetics started developing software and hardware complexes that implemented mathematical models and information technologies developed earlier. One of the first complexes was a dialogue system for alternative economic calculations and macroeconomic forecasting. Currently, components of the system are being improved and its technical support is provided, which is based on the network of modern computers, special means of input and presentation of information, multiprocessor systems (clusters) SCIT-2 and SCIT-3 developed at the Institute of Cybernetics.

Another software and hardware complex planned to implement these models and information technologies is the information-analytical system for decision support at the state level in complicated socioeconomic situations, which can be considered as a prototype of a situation center.

4.2   Econometric Modeling for the Study of Market Transformation

In recent years, the Institute of Cybernetics has improved mathematical models of processes and phenomena in a transition economy. This scientific field is headed by corresponding member of NASU Mikhail Mikhalevich, who is a worthy successor of his father. His research focused on modeling the interaction of competitive segments of the national market and segments with different forms of imperfect competition. Considering the prerequisites for imperfect competition in labor markets under transformation conditions, he was the first to develop and test the following macro models of economic dynamics.

Consider an economic system where competitive market of goods and services and a market with imperfect competition interact. Let the aggregate demand $V(t)$ in such a system be known for each continuously varying instant of time t. Denote by $G(t)$ the total employment and by $w(t)$ the average wage at the time t. Assume that prices $p(t)$ for goods and services depend on the difference R between demand and supply, which, in turn, depends on prices, employment, and time. As a result, we get a system of differential equations that describes the variation of the above economic parameters:

$$\frac{dV}{dt} = a_1 Gw + a_2 V,$$
$$\frac{dp}{dt} = \lambda(V - R(p,\ G,\ t)), \tag{4.9}$$

where a_1 and a_2 are some predefined constants and the function $R(p, G, t)$ is assumed known.

The differential equations that describe the dynamics of the variables G and w are derived for each type of imperfect competition. System (4.9) is supplemented with these equations and with Cauchy initial conditions. As a result, we have a system of nonlinear differential equations that cannot be solved analytically but can be qualitatively analyzed, as was done for model (4.2), (4.3), (4.4), (4.5), (4.6), (4.7), and (4.8). With such an approach, it was shown in [9] that the interaction of the competitive market of goods and services with the monopsony labor market may cause business cycles similar to the classic one. These cycles are characterized by a surge of inflation before the recession ("inflationary boom"), economic depression because of inflationary devaluation of savings, and significant impact of deflationary growth of assets at the stage of economic recovery. Similar results were obtained in [215] for an oligopsony labor market, where there are several employers with significant demands for labor. These employers are assumed to make products of different types, so they do not compete in the market of goods and services.

The paper [7] analyzes the economic dynamics in the case of two-sided monopolistic competition in the labor market. It is shown that under these conditions, there may occur business cycles of postclassical type, that is, without pronounced stages of inflationary boom and depression, with price variation weakly influencing the cyclicity, and with decaying cycle amplitude. The paper [8] continues these studies and considers a model with increasing labor productivity. This model was shown to have Schumpeter's cycles, where cycle stages preserve the general tendency of economic growth. The amplitude of such cycles decreases and, at a certain instant of time, the system begins to develop according to a "balanced growth" scenario.

The paper [10] analyzes economic dynamics models for systems with open labor market with imperfect competition. Such models were considered because the assumption on monopsony (employer's monopoly) or oligopsony (a limited number of large employers) at the macro level may meet an objection since there are many legally independent enterprises under the conditions of a transition economy. At the same time, the assumption on these forms of imperfect competition at the level of individual localities and regions seems to be quite reasonable, since under the conditions of planned economy, with its excessive centralization and concentration of production, numerous monofunctional urban centers appeared, where almost all people worked at one or several related enterprises, and highly specialized regions. According to [195], up to 60% of the population of some Russian regions lives in monofunctional towns; it is 26% for the whole country. At the same time, only 16 major Russian corporations control (directly or indirectly) the employment in 53% of monofunctional towns.

About one third of Ukrainian towns and almost half of urban settlements are also monofunctional [166]. They are mostly concentrated in regions with developed mining (primarily coal), metal, and chemical industries, where the influence of centralized financial–industrial groups is strong. Thus, there is every reason to consider the labor market of such towns and regions as a market with the overwhelming power of employers who are able to influence wages in accordance with their interests. However, these market segments are open since labor migration from (or to) cities and regions remains possible. This was taken into account in the development and analysis of dynamic macromodels similar to those considered earlier.

The analysis of these models showed that cycles of classical and postclassical type (depending on the form of imperfect competition in the labor market) can occur in these systems. The wages at the growth and boom stage in such a system may exceed wages in the interacting sectors with competitive labor market. The inflow of labor force caused by this excess may rapidly reduce its cost at the end of the boom, which will promote the beginning of the stage of decrease in the classical business cycle. At the same time, the outflow of labor force at the stage of depression may cause its shortage and increase in wages, which will initiate business recovery.

Thus, migration process in these models influences the economic dynamics as well as price variation under monopsony and oligopsony and reinforces production cyclicity.

The investigations made it possible to conclude that different forms of one-sided imperfect competition in the labor market adversely affect the economic dynamics. These results can be used to decision support in employment and remuneration policy and to evaluate and select alternatives of the socioeconomic development of the country.

Along with modeling the real sector of transition economy, models of financial processes being crucial for the economy as a whole were developed. The "budget" modeling complex was created. It allowed analyzing the mutual influence of budgetary policy and monetary processes in a state with transition economy. The studies on economic and mathematical modeling were not restricted only to the fiscal, credit, and financial systems. Much attention was paid to the real economy, its interaction with the sector of money circulation. The further development of this system is to take into account the interaction between the central and regional budgets, the impact of monetary and fiscal policies on the economic development of regions, modeling and optimization of the dynamics of external debt payments, and a number of other important aspects of the budget process.

Noteworthy is the fact that the structure of the modeling complex and calculation procedures based on individual models included in the complex experiences significant changes over time due to new taxes, modified calculation procedures based on the current taxes, changes in other regulations and the practice of their implementation, etc. As a result, models will become more complicated due to an increasing number of causal relationships among economic phenomena. Therefore, there arises a problem of creating a universal methodology to efficiently provide the flexibility of the modeling tools being used.

Of great importance for its solution is widespread adoption of methods of cybernetics as a science that studies the general characteristics of complex systems and methods to control them. The fundamental concepts such as the structure of a system, the hierarchy of its subsystems, feedback, and self-regulation become especially important in studying the transition economy and identifying the necessary structural changes in economic relations.

In view of foregoing, the Institute of Cybernetics pays much attention to the development of intelligent information technologies that would combine the processes of modeling (model set up and analysis), choosing an algorithm for model calculations, the convergence analysis of interactive procedures of management decision making and the stability of the results obtained under certain variation of external conditions, etc. These technologies are widely used to solve various problems that need economic, social, and engineering aspects to be considered simultaneously. In what follows, we will deal with a class of such problems.

The results obtained at the Institute of Cybernetics in modeling the transition economy and the tools used are detailed in the monograph [124].

4.3   Approaches to the Creation of Information Technologies for Energy Saving in the Economy of Ukraine

Energy problems such as increase in the shortage of energy resources while the approaches to economic growth remain the same, strengthening of the influence of noneconomic factors on the world's energy markets, the necessity of minimizing the adverse effect of the energy industry on the environment and the need for substantial increase in the efficiency of the available resources, and use of new energy resources have recently become of global importance. States with different levels of development of economy and society regard the solution of these problems as a prerequisite for the mitigation of the consequences of the global financial crisis and transition to sustainable economic growth. Energy saving plays an important role; most experts believe that it can only be achieved through systematic implementation of a number of economic, technological, sociopolitical, and other measures. Ukraine, as a part of the global economy, cannot stay away from these problems, which, being common to most countries, are quite specific in Ukraine. This is primarily due to incomplete structural reforms.

Extremely inefficient use of energy resources has long been a chronic disadvantage of the Ukrainian economy. According to the National Energy Strategy, the energy intensity of Ukrainian GDP is greater by a factor of 2.6 than the world's average level and is 0.89 kg of standard fuel per 1 USD of the GDP. This level is 0.20 for Austria and Denmark, 0.24 for Japan, 0.40 for Finland, 0.43 for Czech Republic, and 0.50 for Belarus. At the same time, the specific consumption of electric energy in Ukraine significantly lags behind the world's leading countries (Fig. 4.1).

Under these conditions, the recovery of economic growth is possible only when it is accompanied by energy saving achieved by overcoming the significant

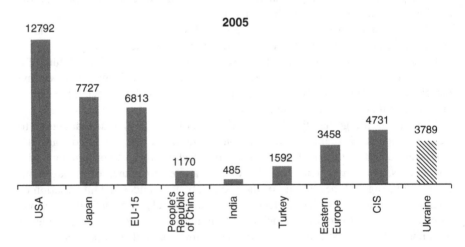

Fig. 4.1 Specific annual electricity consumption in the world and in Ukraine, kWh per person (according to the IEA)

technological lag in most economic branches and in housing and communal services, improving the brunch structure of the national economy in general and export/import operations in particular, and reducing the influence of shadow economy. To enhance energy saving, it is necessary to solve several critical problems of the energy sector of our country. Let us briefly characterize some of them.

The dependence of Ukraine on the import of organic, primarily hydrocarbon fuel (oil and natural gas) is most commonly referred to in scientific journals and mass media. Actually, this dependence is not critical. According to the data presented in the National Energy Strategy of Ukraine [44], our country imports about 60.7% of necessary energy resources, including conventionally primary nuclear energy (i.e., nuclear fuel processed overseas). For comparison, this parameter for France, which has a closed nuclear fuel cycle and an extensive network of nuclear power plants, is somewhat greater than 50%. For other developed countries, it is much greater: 61.4% in Germany, 64.7% in Austria, about 82% in Italy, and more than 93% in Japan. The problem for Ukraine is not the volume of energy import but no diversification of their purchase. Energy supply from Russia (directly or via Russian companies) exceeds 85% of the total energy import, which creates serious economic and political problems. Improved management, access to new foreign energy markets, and diversification of the energy sources are the main ways to solve these problems. One should take into account possible deterioration in the terms of trade (e.g., forced transition to world average prices) and the need for restructuring the specialized transportation infrastructure that will require time and considerable investment.

Meanwhile, Ukraine has large deposits of solid fuel, especially brown coal. However, its low heat capacity, high content of ash and other contaminants, low processability, and high costs of mining complicate the use of this type of fuel in the system of production technologies, and change in these technologies may cause a radically different situation. Considering the reduction of import dependence, it is potentially convenient to create a closed nuclear fuel cycle; however, numerous investment, innovation, environmental, social, and political consequences should be taken into account. Thus, the influence of critical import of energy can only be reduced by combining the improvement in foreign trade management and radical technological changes.

Another important problem related to the fuel and energy complex (FEC) is the undiversified structure of the consumption of primary energy resources in Ukraine. The share of natural gas in FEC has continuously increased over the past 20 years and was about 41% in 2007 against 22% in Western Europe and 24% in the USA. This state of affairs can be attributed to the fact that after the growth of the energy prices in the late 1990s, it was the only type of fuel that Ukraine bought at prices much lower than the world market ones. However, the share of renewable energy sources in Ukraine is very small (4% vs. 6.1% in the EU countries and 7% in the USA). Nevertheless, the most serious problem for Ukraine is not a lack of energy resources but their extremely inefficient use, which was mentioned above. Thus, radical technological changes accompanied by improved management are also necessary in fields related to FEC.

According to estimates given in the Energy Strategy of Ukraine, energy-saving measures will reduce energy intensity of the GDP by 25.8% in 2010, by 37.3% in 2015, and by 44.1% in 2020 compared to the current level. Clearly, this cannot be achieved without a scientifically justified comprehensive program of fundamental structural and technological changes aimed at energy saving. To develop such a program and to monitor its implementation, an appropriate tool should be developed.

A series of studies aimed at the elaboration of modern information technologies for decision support in the development of fuel and energy complex and intensive energy economy in Ukraine have been carried out at the Institute of Cybernetics within the framework of the comprehensive research program of the NAS of Ukraine "scientific and technological basis for energy saving" ("energy saving" for short). M. V. Mikhalevich, corresponding member of the NAS of Ukraine, performed overall scientific management of these studies. The purpose of the research conducted jointly with the Institute of General Energy of the NAS of Ukraine was also to use these technologies to elaborate recommendations for improving the situation in the fuel and energy sector, the following aspects being emphasized first of all.

Energy saving in the broad sense, that is, reducing specific consumption of energy and of all kinds of power resources, is one of the most important tasks of modern economic development. As Academician Dolinskii noted, "... energy saving and energy consumption efficiency should be considered as an important additional power resource for Ukraine, no less important than oil and gas" [36]. This is due to a number of factors.

First, the current Ukrainian system of industrial technologies has formed for decades under conditions that met neither the requirements of market relations nor the realities of modern business activity. These conditions were:

- Low prices for natural raw materials (especially power resources), which did not help saving them
- Low prices for basic consumer goods at low wages, which caused a small share of value added in the price of most products
- Developed system of budgetary subsidies, which distorted the interbranch price ratios and artificially reduced prices for one goods and increased them for others
- The closed nature of socialist economy, which did not promote the competitiveness of domestic goods in foreign markets

Currently, the prices for most types of power resources imported by Ukraine have reached the world level and often exceed it because of imperfect production technologies. This leads to a rapid increase in material costs in the basic industries (iron and steel, metal-intensive machinery, chemical, construction materials, agricultural, etc.). Products of these industries remain competitive mainly due to low remuneration of labor and shadow schemes used to minimize fiscal costs, which poses serious threats to social stability and national security.

Since this was understood, "The Conceptual Framework of National Policy on Efficient Use of Fuel and Power Resources (Energy Efficiency)" put in force by the decision of the National Security and Defense Council of Ukraine on May 30, 2008 stated that "... under intensified global competition for fuel and energy resources, the energy efficiency of the economy and, respectively, reduced cost of energy resources for production is the only way to obtain competitive advantages of Ukraine in global markets and welfare of citizens. Ensuring the competitiveness of the economy of Ukraine by currently popular methods, namely, by reducing wages, refusing from long-term investments to modernize production facilities, involving tax avoidance schemes, etc., are unacceptable for the development of the state. Such approaches do not improve the fuel and energy efficiency of the national economy, maintain high energy intensity of the GDP of Ukraine and the consumption of scarce energy resources, which threatens the national security of Ukraine in the economic, power, environmental, and social fields" [147].

Other effects of the above methods are reduced effective demand and reduced share of the value added in the price of products and development of Keynesian macroeconomic processes, where the price weakly depends on the demand for manufactured products and the production volume on the price. One of two alternative scenarios is implemented under these conditions, depending on how tight the current monetary policy is: growth of nonpayments and intensified decline in production through the reduction of final demand or accelerated cost inflation. A way out of this situation is to increase the solvency of consumers (primarily of the population) and, simultaneously, to reduce the production costs.

Second, no less dangerous than the uptrend of production costs is stronger dependence of the economy of Ukraine on the import of certain types of energy resources. This primarily applies to natural gas. Its price and supply become mainly political issues. One should not exaggerate the impact of this factor: For many industries, technologies allow replacing gas by other fuels; however, this increases material costs and environmental pressure. The exceptions are certain subsectors of chemical industry (ammonia, carbamide, and nitrogenous fertilizer production and synthesis of some polymers) where gas (methane) is used as a raw material for chemical reactions and, thus, is irreplaceable. However, given the aforementioned adverse effects of increased production costs, it is reasonable to consider the reduced consumption of natural gas under the conditions of stabilizing economy as one of the most important tasks of structural and technological transformations.

Third, the current condition of equipment in the major industries, especially in nonproduction sphere (e.g., housing and utilities) requires immediate significant investments to renovation. The main source of these investments should be the national income generated within the country, which is a part of the value added of manufactured products. Under these conditions, an increase in the share of value added in the price of manufactured products should be considered as one of the most important in overcoming the economic crisis and renewing the economic development.

Note that the above problems (reducing tangible costs and natural gas consumption and recovering investments in production) are interrelated since one of the major ways of their solution is energy saving; however, they are also contradictory to some extent. Different energy-saving measures differently affect the achievement of these goals.

For example, replacing liquid fuels with natural gas bought at a reduced price decreased overall tangible costs (in monetary terms, at current prices) but stimulated the rapid growth in gas consumption. Closing of chemical plants being the major (in addition, having no alternative) gas consumers will reduce the need for it but will also reduce gross value added, employment, and investment resources, that is, will aggravate the crisis in the economy of Ukraine. Thus, making energy-saving decisions, one should simultaneously solve all these three problems, which should be reflected in the list of criteria used to evaluate decisions on FEC reforming. One should take into account that the situation under which the reform will be implemented is unpredictable and there are many unpredictable factors, both external (world prices for energy resources, global market conditions in traditional Ukrainian exports) and domestic (political situation, state of the financial system, etc.). Thus, decision making in energy saving and FEC reform are multicriteria problems subject to numerous informal aspects, and the decisions are made under risk and uncertainty.

In characterizing the ways to solve energy-saving problems, we should note that they may be both direct and indirect. The former include measures to reduce specific (i.e., per unit of product) consumption of energy and all types of energy resources in production technologies and to introduce new technologies with significantly lower costs. Of great importance is reducing the energy consumption during the transmission and more effective use of available energy resources. Direct ways of energy saving also include the use of alternative energy technologies, new energy resources (agricultural and forestry waste, ethanol-based motor fuel, biogas, syngas, and low-octane gasoline – gas oil, produced from brown coal, etc.). One should bear in mind that introducing these measures needs significant investment resources and often increases (at least temporarily) tangible costs. Therefore, these measures make sense due primarily to the criterion of minimizing the cost of imported energy resources such as natural gas.

Indirect energy-saving measures such as reducing the specific consumption of most energy-intensive products, replacing domestic production of such products with import, etc. have no less, if not greater, potential than direct measures have. For example, replacing metal components with composite ones in manufacturing vehicles and machinery will reduce the need of manufacturing engineering in products of ferrous metallurgy, which is energy-intensive production. If the production of the necessary amount of composite materials does not increase power consumption, the total need for energy will decrease due to reduced production of metals. Improving the reliability of engineering products will have a similar effect; it will reduce the need for spare parts for repair and hence the overall energy consumption due to reduced production of these parts.

There are more proposals for indirect energy saving than for direct one. Implementing each of the proposals may exert both positive and negative influence: for example, increase the energy-intensive production of some kinds of nonferrous metals to be used then to manufacture composites and increase production costs due to installing and developing new equipment. Comparing the positive and negative consequences of these proposals requires a comprehensive and profound economic research using quantitative methods, including economic and mathematical modeling. Such models, together with the above-mentioned decision support methods, form an important component of information technologies developed at the Institute of Cybernetics, which is a recognized research center for developing methods of discrete, stochastic, and multicriteria optimization, computer modeling of complex systems, processing of expert judgments, and other means of modeling complex systems.

These models reflect both technological and economic aspects of energy-saving measures. The latter produce a significant effect in their implementation at the level of individual large groups of related industries, at the branch and interbranch levels, which determines the degree of model aggregation.

Based on these objectives and ways of energy saving, the main attention in the research was paid to the following problems, which will be solved by the proposed tools:

- Identifying the most effective ways of energy saving, taking into account the specific features of the national economy
- Improving the efficiency of investment and innovation energy-saving measures
- Accounting for foreign economic aspects, taking advantage of the international division of labor
- Improving the market relations between energy suppliers and consumers to encourage energy saving

With due regard for these problems, the main efforts in the research are concentrated on the development of economic and mathematical models of four types, which have a common information base and, together with the necessary software and algorithms and developed user interface, are the core of a specialized decision support system.

The *first group* includes models of planning the main aspects of structural and technological changes, which consider, in particular, reduction of specific production costs due to improving available technologies and introducing new ones. Along with reducing energy consumption, this reduction can increase, without inflation, the share of value added in product price, which creates prerequisites for the growth of wages and other incomes of ultimate consumers. This problem can be solved using input–output models with variable input–output coefficients, which determine the volumes of specific consumption of different products using different technologies. The modification of production technologies necessarily changes these coefficients; therefore, to determine the ways of best modification, one should assess the preferred directions of such changes. This can be done with the help of nonlinear optimization models.

Because the purposes of energy saving and reduction of overall tangible costs are not identical (though correlate with each other), it is important to consider multicriteria optimization versions of the developed models. Calculations with the use of these models involve solving complex nondifferential optimization problems with nonconvex objective functions and sets of feasible solutions. Under these conditions, single use of local search optimization methods for model calculations cannot answer the question whether the set of constraints is compatible. This problem can be solved using RESTART technologies, which involve repeated calculations from different initial points. These technologies can be relatively easily parallelized using multiprocessor computer systems, which find increasingly wider application. To this end, SCIT-2 and SCIT-3 clusters were used.

Note that the results of calculations based on the proposed optimization models under the conditions of transition economy cannot be prescriptive. They should help to identify the desired structure of production technologies that would intensify the socioeconomic development of the country, identify the ways to convert the current structure into a desired one, to evaluate the necessary resources, etc. Of importance is an analysis of the time variation of coefficients of direct costs over past few years to ascertain whether their actual values tend to or diverge from the desired values obtained using models. From this point of view, obtaining a series of problem solutions using nearly optimal values of criteria seems to be more reasonable for further application than searching for the optimal solution.

The models under consideration include a number of ambiguous parameters. Therefore, it is expedient to carry out interactive calculations for different values of the parameters. It is desirable that the interval between the beginning and the end of a series of calculations, that is, between generating an alternative model with specific values of its parameters and obtaining a series of its solutions, would not exceed at least 3–5 min. For real high-dimensional problems, it is not always possible to satisfy these time limits with one-processor computers. However, this can be achieved rather easily with multiprocessor systems such as the clusters mentioned above since the time it takes to carry out the entire set of calculations with the use of RESTART technologies, after paralleling the computing procedure, will be equal to the time of single implementation of the computational algorithm for solving one problem. This approach was implemented in a specialized computer system.

Model calculations with real data were carried out using such a system. The results of modeling allow identifying types of production costs whose reduction may have the strongest effect on the total energy consumption under the conditions of increased (due to energy saving) supply of and demand for goods of ultimate consumption. The calculations were based on recent intersectoral balances (for 38 branches); for comparison, similar calculations were performed for the balances of the mid-1990s (for 18 branches).

Input–output coefficients (which are expedient to reduce according to the calculations based on recent balances) are mostly related to energy costs and construction materials in the basic branches of economy of Ukraine. In particular, it is reasonable to reduce, by 47–50%, unit costs of coal and hydrocarbons (especially natural gas), products of chemical-recovery coal carbonization, and power industry in metallurgy (e.g., by increasing the share of metal melted from scrap and applying modern electrometallurgy technologies). Specific consumption of hydrocarbons and electricity should be reduced by 45–48% in mechanical engineering and by 43–46% in chemical production. Specific consumption of hydrocarbons in the oil refining industry should be reduced by 49% and specific consumption of coal in the coke industry by 50%. It is expedient to reduce the specific consumption of coal, hydrocarbons (mostly natural gas), and production of coke industry in the production of energy by 46–48%. The specific consumption of products of oil refining industry in agriculture, construction, and transport should be reduced by 49–50%.

It is also expedient for energy saving to reduce the consumption of construction materials whose production requires much energy. In particular, the consumption of metal products in mechanical engineering (one of the ways is to reduce the share of obsolete technologies of metal cutting), for mining, and in construction should be reduced by 48–50%. It is recommended to reduce the specific consumption of products of the industry of nonenergy mining (iron and manganese ores, fluxes, fire clay, building stone, and sand), which is also 48–49% in the steel industry and construction. As calculations showed, auxiliary costs of production of key branches are overstated. Such costs can be reduced by 45–47% in metallurgy, in mechanical engineering, and in oil refining and chemical industry and by 36–38% in the by-product coking industry and in the food industry. For some industries, it is expedient to reduce the specific consumption of mechanical-engineering products, which could be indicative of inefficient use of the available equipment. This applies to agriculture (49% reduction recommended), mining (48% reduction), and electric power industry (46% reduction). As compared with the results of calculations for the 18-branch balance, the number of items associated with auxiliary costs to be reduced significantly decreased. Transportation costs remain too large in construction (reduction by 48% is recommended), and specific (per unit of manufactured products) services of financial intermediaries and other entities in engineering and services of gas supply in housing are expedient to be reduced by 46–49% and 18%, respectively. Given the results of calculations according to the 18-branch model, a significant reduction of some input–output coefficients whose values were "critical" is indicative of some improvements over the last decade. This includes the expenses of own products in the coal mining industry, which have reduced by more than 30% for this time. However, reducing this factor remains desirable (the recommended reduction is 24%). At the same time, recommendations as to reducing the specific consumption of energy resources and construction materials remained almost the same for both 18-branch and 38-branch models. This testifies that unfortunately, there were no real changes in energy saving through technological renovation of production in Ukraine during this time.

The above reduction in material costs will increase the share of wages in the price of production of most sectors from 3–7% to 15–25%. This recommended increase differs for different branches: 10–17% in the agriculture, light and food industries, and mechanical engineering and up to 3–4% in the chemical, petrochemical, forestry, and woodworking industries.

Alternative solutions obtained using RESTART technologies differ insignificantly in their structure. Components of the solutions that differ significantly pertain to relatively insignificant fields. Recommendations as to reducing input–output coefficients in the fields such as metallurgy, mechanical engineering, chemical and coke and by-product process, oil refining, electric power, food, and transportation industries are the same for all the alternatives.

Implementing the proposed changes requires investments of about 45 billion US dollars. Considering only the immediate effect of these changes, that is, the increase of final consumers' income by 59.1 billion of US dollars, the efficiency of investments is more than 30%, which is rather high. Furthermore, only reducing the above specific consumption of natural gas in industry and consumption of energy-intensive production will allow annual savings up to 14.3 billion cubic meters. Reducing the need for energy-intensive products will also provide annual saving of about 7.8 million tons of coal and 5.5 million tons of steel. The potential GDP will grow by 25–30%. Thus, energy saving and structural and technological transformations have to become an essential factor in overcoming the economic crisis. Considering the urgency of energy saving under the current conditions in Ukraine, it is expedient for public authorities to allow for the results of the calculations in forming the budgetary (when prioritizing public investment), industrial, and overall economic policy. Noteworthy is that the modeling system created by cyberneticians is open. It can take into account the emergence of new goals and change of priorities among existing ones and new restrictions and tools to achieve goals; the system can include new analytical procedures.

Let us consider another aspect of energy saving. A large number of innovations are now developed, both purely technological and organizational, aimed at reducing specific energy consumption and reducing power-consuming productions in different economic branches. Each of the proposed innovations is characterized by estimates of changes in production costs such as energy consumption and by possible increase in the final incomes of producers due to its implementation. The substantiation of innovation should include the assessment of the need for investment, financial, material, and other resources for its implementation. The scarcity of these resources does not allow implementing all the proposed innovations, which gives rise to the problem of their selection. This task is complex and multidimensional; it cannot be completely formalized. Nevertheless, applying mathematical models and numerical algorithms for calculation is an effective means of decision support in those issues.

Thus, the *second group* of models that were developed during the research project was models of selecting energy efficiency projects. These models can be used to select technology projects aimed, for example, at increasing the level of final consumption, reducing the consumption of scarce energy resources, improving the technological structure of production, and stimulating the scientific and technological progress. In selecting projects, the scantiness of investment resources and various technological, environmental, and other restrictions should be taken into account. Complicated multicriteria optimization problems but with discrete variables rather than continuous variables (as for the first group of models) arise as well. This imparts specific features to the development of computer algorithms based on descent vector method, simulated annealing, and tabu search and necessitates the use of multiprocessor computers.

Project selection models differ essentially from the previously mentioned models of planning structural and technological changes. They are no longer the means of assessing general trends in technological development and identifying the main ways of energy saving but are the basis for decision support of specific managerial decisions on the (in)admissibility of certain innovative energy-saving proposals. All this has an effect on the calculation with models.

For the second group of models (as well as for the first one), it is important to conduct interactive calculations to account for unstructured aspects of decision making; however, the subject of the dialogue between the modeling system and decision makers is somewhat different. In particular, allowance is made for the possibility of adopting specific projects, based on informal reasoning and partitioning the set of all the projects into several subsets with different priorities, which is also established informally. The projects are first selected from a subset of higher priority and then, within the remaining resources, from a subset of lower priority. From this point of view, it is of interest to develop a problem-solving procedure, primarily based on local search methods. The iterative nature of these methods allows implementing this approach. A significant amount of computation required to solve the problem of project selection in this way creates certain difficulties for the implementation of interactive procedures. Intervals between successive references to the decision maker during the dialogue procedure can be substantially reduced by applying multiprocessor computers. Local search algorithms, both deterministic (descent vector method) and stochastic (simulated annealing and limit equilibrium search), provide a relatively simple and effective parallelization. Unlike calculation methods that use models of planning of structural and technological changes, such paralleling will not be based on input data but directly on the computational scheme of the algorithm. This applies, first of all, to the analysis of individual subsets of the unit neighborhood of the current approximation to the optimal solution on separate processors. The best alternative changes found with independent processors are later compared by the coordinating processor, and the next approximation is finally selected.

To carry out calculations using the project selection models, a software system compatible with the system of planning of structural and technological changes in the requirements for input data and principles of user interface was developed at the Institute of Cybernetics. The latter is based on a hierarchical menu system. Based on informal reasoning and using special windows, the user may establish the need for unconditional acceptance of some of the projects and priorities among the remaining projects. The project selection problem is then solved for the set of remaining projects and for the volume of resources that remains after the implementation of preselected projects. Means for the analysis of the results are provided. To this end, the positions of the matrix of direct costs that are reduced or increased after the simultaneous implementation of all selected (both based on informal considerations and model calculations) projects are colored. This allows comparing the results with the general recommendations for structural and technological changes obtained by previously described models.

Implementing energy-saving measures usually have a favorable effect on the environment because reducing energy consumption is accompanied by the decrease of its production and hence the reduction of the associated pollutant emissions. On the other hand, introducing new, more economic (according to the current system of prices and fiscal policy) technologies of energy production, transmission, and use can be accompanied by unpredictable impact on the environment, especially in the long run. In our opinion, an analogy with nuclear energy is appropriate since the development of this technology at the beginning of its introduction was associated for many experts not only with saving money for buying organic fuel but also with reducing the adverse environmental impact. The awareness of the problems that arise during the storage and disposal of radioactive waste, the impact of possible major disasters, and the consequences of low doses of radiation came much later. Thus, evaluating all the aspects of the environmental impact of energy-saving projects, we are dealing with typical problems of decision making under risk and uncertainty. The possibility of applying both deterministic and purely probabilistic approaches is restricted by a number of factors (unlikely events with very significant consequences, heterogeneity of statistical samples, the influence of different types of uncertainty, etc.).

Over the last years, scientists of the Institute of Cybernetics have created mathematical tools for decision support under the above conditions. Noteworthy are methods for processing weakly structured statistical information (the GMDH method and methods of probability estimation and statistical modeling of rare events), quasiprobability approaches to decision making (e.g., fuzzy set methods), Bayesian recognition and classification procedures, and algorithms for decision making under risk and uncertainty (stochastic optimization, scenario approach, etc.). All these tools are planned to be used for further development of project selection system, primarily to account for long-term environmental and other consequences of projects. The risk and uncertainty factors are also significant for the next field of research.

The *third group* of developed models is the models of determining a rational branch structure of domestic production, export, and import in order to make better use of the advantages of the global division of labor, which is reflected in world market conjuncture (or models of changes in the structure). Since it is also necessary to consider various target values such as export–import balance, domestic consumption, and energy costs, multicriteria cases of these models are of interest. The profound influence of risk factors and uncertainty necessitates the development of stochastic models, such as two-stage ones, in which a preliminary decision is then adjusted after the refinement of the data concerning uncertain model parameters.

Improving the branch structure of export and import is one of the most important ways of energy saving. This problem is also closely related to structural and technological transformations. On the other hand, such transformations are effective only if they are accompanied by a shift of emphasis in foreign economic activity.

Imperfect production technologies, their great energy and resource intensity, and low quality of consumer goods make the most of the domestic products of final consumption noncompetitive in world markets. Lack of competition both in the domestic market (where direct or indirect protectionism takes place) and in the foreign markets (where our businessmen who manufacture end-use products often even not try to enter) does not stimulate improving the technical level of production and even increases the technological gap. The competitiveness of products keeps reducing. As a result, we now have well-developed industry, which is unable of producing final consumption products competitive in the world market. The recent structure of Ukrainian export has been very nondiversified and consisted mostly of energy-intensive intermediate-consumption products (metals, chemical intermediates, etc.), whose production depends on the import of certain energy resources (primarily natural gas) and sales is concentrated in limited foreign markets. It is no wonder that this structure appeared to be extremely vulnerable to the precrisis growth in prices for energy resources and to the sudden reduction in demand for investment goods, which always accompanied the beginning of global financial crises.

Under these conditions, the use of the advantages of the global division of labor and deeper integration of the country in the system of economic relations should be major tasks of structural reforms.

Attracting foreign investments, especially technological innovations, is crucial for the improvement of industrial base. Joining the global division of labor, producing primarily the products that have benefits, and importing cheaper products allow reducing the share of intermediate consumption in the output aggregate, efficiently using the available resources, and thus improving the performance indices of the economy as a whole. In fact, the use of the advantages of the international division of labor can solve the same problem as the improvement of production technologies. These processes are interrelated; therefore, they should be considered as a system.

The operation of the economy depends heavily on external factors such as the volume of export and import, investment inflow and outflow, and world market fluctuations. This dependence is especially important for countries with open unstable transition economy such as Ukraine and, consequently, whereas it is difficult to achieve sustainable economic growth without external aspects.

Therefore, mathematical models were proposed to find a rational branch and geographical structure of export and import with regard to the structure of production technologies and the volume of domestic consumption, current restrictions for the possible turnover of different types of domestic products in foreign markets and its prices, limited resources and production facilities within the country needed to change the output of certain products, and other factors.

The models were multicriteria and involved expert estimates. Their significant feature was that various risk factors and uncertainties were taken into account. To this end, different approaches were provided: scenario forecasting, probabilistic modeling, stochastic optimization, interval and order expert estimates, etc. All these approaches have been intensively developed all over the world during the recent decades with active participation of leading scientists of the Institute of Cybernetics, and Academician Mikhalevich also initiated the studies in these fields.

The models were supplemented with numerical algorithms and information and software components. They formed a prototype of the simulation system based on the same principles as the complex models for identification of the main ways of structural and technological changes and selection of energy-saving projects mentioned above. This system was used to carry out calculations with recent real data, which lead to the following conclusions.

The branch structure of the export from Ukraine that is the best in the balance of external payments and resistance to possible changes in the world markets is as follows:

- The share of the chemical and metallurgical complex (ferrous and nonferrous metallurgy, chemical, and oil refining and by-product coking industries) should be about 40% of the total export, the share of ferrous metallurgy should not exceed 25%, and the share of the export of structural materials for high-technology industries (composite materials, alloys with special properties, semiconductor raw materials, single crystals, etc.) should be increased to at least 5% of the total export.
- The share of the machine building industry should be about 18% of the export, special attention being paid to promoting such products to the markets of developing countries.
- The share of agricultural products should be about 20% of the export, the diversification of the market outlets being of importance.
- The share of the export of transport services, including the transit of hydrocarbon materials by pipelines, should be about 19% of the export.
- The rest branches of the economy should be about 3% of the export.

Comparing this structure with the one that has prevailed in Ukraine over the past decades, we should pay attention primarily to the very large share (60%) of the production of the chemical and metallurgical complex in the actual export from Ukraine. Its major portion is products of ferrous metallurgy and inorganic chemistry, that is, the sectors that require the maximum amount of imported energy resources. The share of structural materials for high-technology industries is much less than the recommended one. Full use is not made of the export potential of the agroindustrial complex, especially mechanical engineering, where metal-consuming subsectors dominate. Such an irrational structure of the export has made our economy vulnerable to rapid changes in the external conditions and is the major cause of the severe consequences of the global financial crisis for Ukraine.

The recommended structure of import is close to the existing one: Most of it should be raw materials and energy resources (about 68%), industrial equipment and technologies, and expensive and high-quality consumer goods. An important requirement is import diversification: The share of each country (group of countries) for each group of imported goods cannot exceed 30%.

Analyzing these data, we may conclude that the export capacity can be improved by expanding the export and reducing the import of products of mechanical engineering and chemical industry.

Implementing the proposed measures to improve the branch structure of export will reduce the amount of critical import to Ukraine against the current one by 14.3 billion US dollars and reduce the annual industrial consumption of natural gas by 9.5 billion cubic meters.

Calculations made it possible to evaluate the role of the food industry in the export potential of Ukraine. It should be a kind of a damper that would mitigate the fluctuations in the world markets of metallurgical, chemical, and machine engineering products. Moreover, the food industry is less energy intensive. The volume of export of food products should remain almost constant (within 3.5–4 billion US dollars) for all the scenarios of world market conditions considered during the model calculations. With its function taken into account, the diversification of the subsectoral and geographic structure of export in this field becomes of primary importance. In this regard, the participation of Ukraine in international associations such as GUUAM, PABSEC, etc., and further institutional development is important.

Along with planning of energy-saving measures, implementing such measures gains in importance. To assess the impact of the proposed plan, monitor its implementation, adjust it if unforeseen circumstances appear, appropriate tools are also required. In view of this, the researchers of the Institute of Cybernetics NAS of Ukraine created the *fourth group* of models consisting of simulation models of economic dynamics. They incorporate various types of imperfect competition in resource markets.

Noteworthy is that Ukraine inherited a highly monopolized, rigidly centralized system of management. At the first stages of market reforms, monopoly and other types of imperfect competition inherited from the previous period created preconditions for obtaining super profits by certain entities and became a nutritious medium for the "shadow" economy and corruption. The fuel and energy sector is a field where these trends are very common.

Developing an effective national economy provides a legal framework to limit monopoly, prevent unfair competition in business, and state control over the compliance with rules of antitrust laws. Nevertheless, while imperfect competition plays a significant role in the economy, it should be taken into account, but at the same time, measures that would minimize its impact should be planned. The above-mentioned models of the fourth group can be used to assess the effectiveness of such measures. In particular, as modeling demonstrates, it is expedient to pass the laws that would regulate the venture business and engineering activities and work for "free" professions and would encourage individual careers in applied, scientific, and developmental research and creation and implementation of new technologies. Activities in these fields usually require significantly less energy consumption per unit of value added created; therefore, this will not only reduce the negative impact of imperfect competition but also directly contribute to energy saving.

The research conducted at the National Academy of Sciences under the Energy Saving Program resulted in a system of economic and mathematical models schematized in Fig. 4.2, the information technology of decision support in energy efficiency, and energy-saving recommendations obtained based on it. This technology can be applied in the National Agency for Efficient Use of Energy Resources or other state bodies that form the energy policy. The measures proposed will reduce the import of energy resources by 26.1% against the last year's level and will save 23.8 billion cubic meters of natural gas (of which 14.3 billion cubic meters due to

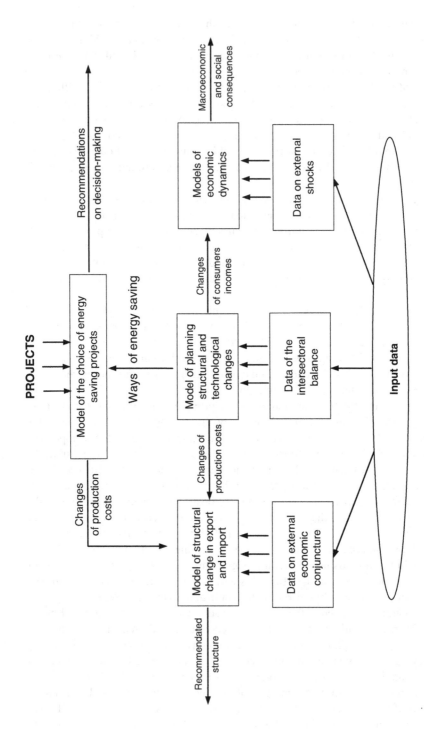

Fig. 4.2 Economic and mathematical models of the Energy Saving Program

reducing specific technological expenses of gas and energy-consuming products and 9.5 billion cubic meters due to improving the structure of export and import), that is, will provide the solution of immediate energy-saving problems determined by the National Energy Strategy. However, the above-mentioned technology and recommendations will be effective only if used on a system basis. Technological changes should be accompanied by structural reforms in all areas of economy and society and should rely upon these reforms. A reform of the fiscal system is necessary to give the manufacturer the most of the additional revenue due to implementing energy-saving measures and in order that subjective tax preferences would cease to be the most affordable way to get super profits. The antitrust regulation should prevent unscrupulous producers' attempts to shift the consequences of rising prices for energy resources to consumers of their products. It is necessary to overcome the shadow economy and corruption, which can absorb additional revenues received due to implementing technological changes and reduce to zero the public interest in such changes. The financial system has to ensure the concentration of resources for structural changes. Thus, energy saving may become a catalyst for badly needed reforms and should be based on these reforms. However, this requires political will and understanding of the fact that improving the economic efficiency as a way out of the current economic crisis has no alternative.

Chapter 5
Analyzing the Solution of Complicated Combinatorial Problems

Abstract Results of investigations of problems of solving complicated combinatorial problems are analyzed. An analysis of problems arising in graph theory occupies a considerable place. Prospects are shown for the combined application of results of graph theory and modern optimization methods to the investigation and solution of complicated mathematical problems.

Since 1960, various organizations have encountered practical problems in planning and design. Solving them required the basic theoretical developments by Mikhalevich and his disciples at the Economic Cybernetics Department.

As already mentioned, the method of sequential analysis of variants resulted from a qualitative leap in research related to the application, improvement, and generalization of linear- and dynamic-programming methods. In problems solved by this method, this fact had an effect on the variables: They were more often continuous rather than discrete (integer-valued). Later, the wide use of both continuous and discrete variables in mathematical formulations of real problems necessitated referring to newest divisions of mathematics such as number theory and combinatorial analysis. On the other hand, computers introduced in various fields of human activity in early 1960s and the success they brought about shifted mathematicians' interests toward discrete methods. Though first combinatorial problems go back 4,000 years, combinatorics dealt mainly with entertaining problems (puzzles, gambling, etc.) for a long time. With the advent of computers, theoretical combinatorics began to develop rapidly and fruitfully and covered various subjects such as enumeration and extremum problems; existence, choice, and arrangement problems; and geometrical and algebraic interpretations. Relevant foreign studies were translated and several Russian books were published. The majority of authors were concerned about two aspects: obvious difficulty of identifying the limits of combinatorics and many various methods and results proposed by combinatorics that are more or less independent of each other. The idea was conceived that combinatorics can be divided into the following large divisions: *enumeration theory,* which covers generating functions, inversion

I.V. Sergienko, *Methods of Optimization and Systems Analysis for Problems of Transcomputational Complexity*, Springer Optimization and Its Applications 72, DOI 10.1007/978-1-4614-4211-0_5, © Springer Science+Business Media New York 2012

theorems, and finite differencing; *order theory,* which considers finite ordered sets, lattices, matroids, and existence theorems; and *configuration theory,* which includes graph theory, flowcharts, substitution groups, and coding theory. The first two divisions drew the attention of mainly theoreticians. The third division was of interest owing to its practical importance; implementing the results required more powerful computers. As already mentioned, Mikhalevich succeeded in materializing Glushkov's last plan to develop a macroconveyor computer system (MCCS) that would be based on the principles of parallel computing and the internal mathematics of a computer.

The main stimulus for the development of parallel-computing equipment has been the need to improve the efficiency of solving formidable problems. The efficiency of parallel computing depends on the computer performance, memory size and structure, and throughput of data channels. However, it also depends at least as much (if not more so) on the level of development of programming languages, compilers, operating systems, numerical methods, and many accompanying mathematical studies.

V. S. Mikhalevich mobilized the efforts of the Institute of Cybernetics and explored every avenue to overcome the difficulties and successfully complete the creation of a macroconveyor system. The majority of problems that arose in the development of the system can be classed among three groups. The first group represents the case where a parallel computer does not equally process all information flows. Considering the architecture of a computer, it is very important to understand what data flows are and to describe them mathematically. The second group is determined by the relations among the operations of a computer program. And, finally, the third group includes problems related to the choice of a coding technology. All this should be considered first of all when paralleling a program, allocating memory, and organizing data transmission.

The colleagues and disciples of Mikhalevich were repeatedly convinced that many problems related to parallel computing performed by different methods had much in common. These problems should be thoroughly analyzed and generalized to find relations among them and ways to solve them. Clearly, such relations are not obvious and can only be found based on rigorous mathematics. Considering specific features of parallel computing allowed concluding that studies in graph theory and combinatorial analysis can be a basis for the successful solution of these problems. These studies directed by V. S. Mikhalevich produced weighty theoretical and practical results and are now continued at the V. M. Glushkov Institute of Cybernetics under the supervision of G. A. Donets who became the head of the Economic Cybernetics Department in 1994. Let us discuss the most influential achievements in solving a number of important problems in the field of graph theory and combinatorics. We will also present some substantial statements of important practical problems whose solution gave impetus to weighty generalizations and development of applications for the method of sequential analysis of variants and other mathematical optimization methods.

5.1 Applying Optimization Methods to Solve Economic Problems

The method of sequential analysis of variants was first tested in solving problems of construction. In 1961, V. I. Rybal'skii, deputy director of the Research Institute of Construction Production (RICP) of Derzhbud of the UkrSSR, offered Mikhalevich to cooperate in elaborating an optimal plan for constructing and launching several thermal power plants in Ukraine. This plan was a part of the long-term national economy development plan issued by the State Planning Committee (Derzhplan) of Ukraine. It was planned to construct 24 new 2,400 MW thermal power plants in Ukraine over the period from 1962 to 1980. When developing a mathematical model of the problem posed, it was necessary to take into account some features of the construction process. It became obvious from the very beginning that the available linear- and dynamic-programming methods were unable to solve this problem, and nonlinear-programming methods were not developed well at that time. One more feature was that the problem formulation disguised unequal operating conditions for the plants to be constructed and for the construction organizations. In particular, the problem formulation was required to take into account both rhythmic operation of future plants and constant workload for the construction and erection organizations, that is, a *continuous-flow* operation.

This condition could not but had a serious effect on the total cost. Applying flow methods increased the work output. Moreover, different objects grouped in flows were affected by different terms of their construction because of different overhead costs, mechanization cost, creating industrial bases, temporary constructions, transferring building organizations from one place to another. These factors had a strong effect on the total capital investments for both individual enterprises and the whole set of enterprises. Nevertheless, these factors are not all equivalent. A great number of workers usually participate in power plant construction, much greater than during the operation of the plant. At the same time, large power plants are as a rule constructed outside cities and settlements. Therefore, it was impossible to accommodate all construction workers in the available housing or a permanent settlement for maintenance staff. A settlement had to be built and then dismounted, after the construction was complete. Taking into account this feature of power construction, the USSR Ministry for Power Plant Construction employed reusable demountable houses. However, up to 40 % of the cost of temporary settlement was not returned, which had a strong effect on the total cost.

As a rule, large expenses went into constructing temporary buildings and structures. Eventually, this practice was canceled, and the majority of construction manufacturing processes were transferred to regional building-industry bases, and installation works were mainly carried out at building sites. This reduced the costs related to temporary buildings and constructions at building sites; nevertheless, the role of the costs of creating regional bases of the building industry increased abruptly. At the same time, there were cases where a long-term plan of construction was composed for existing regional bases, that is, their creation did not depend at all

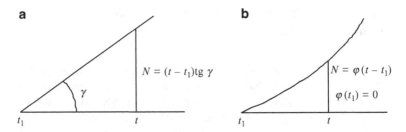

Fig. 5.1 Increase in the number of workers: (**a**) for a constant output, (**b**) with regard to increasing output

on the grouping of construction objects, and thus, there was no need to take into account the relevant costs. It is such a case that was dealt with when planning the construction of large thermal power plants in 1962–1980 since according to the data taken from documents of the USSR Ministry for Power Plant Construction, creating relevant regional bases in Ukraine should be completed till 1965, that is, to the beginning of the development of the first continuous construction flows.

Thus, in each specific case, it was necessary to clearly understand the value of one factor or another during calculations, and then to compose a general algorithm to calculate optimal terms and sequence of the construction. Based on typical calculations for the construction of a thermal power plant (such calculations were then performed at the RICP of Derzhbud of the UkrSSR), relevant software can be used to find optimal parameters for the construction of one power plant and optimal parameters for the construction of several such power plants grouped into one continuous flow.

One of the key parameters in terms of which the other parameters are expressed when power plants are constructed as a continuous flow is the rate of increase in the number of workers. This number grows from zero to the necessary level, remains constant for a certain time, and then decreases to zero because the workers and the equipment are transferred to the construction site for the next power plant in the flow. All the typical calculations at the RICP were performed for a constant labor productivity (or output). The time dependence of the number of workers was linear (Fig. 5.1a).

At the Institute of Cybernetics of the Academy of Sciences of the UkrSSR this function was proposed to be generalized to the case of increasing labor productivity and to be represented as $\varphi(t)$ (Fig. 5.1b). Then the scope of construction and erection works for the whole flow can be defined as the area bounded by the curves of the number of workers at each plant during the construction period T and by two straight lines $y = N_{max}$ and $y = 0$ (Fig. 5.2). To calculate the scope of construction and erection works, integration formulas were applied, which necessitated standard calculations for individual power plants and flows.

The long-term plan stipulated that all power plants should be constructed independently, with no coordination with each other. The calculations that had to be carried out according to the developed algorithm and the program at the Institute

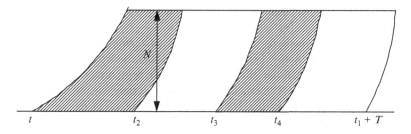

Fig. 5.2 Need in workers to construct four power plants (as output grows)

of Cybernetics determined the advantages of the continuous-flow method compared with the long-term plan. The sites and capacity of the plants were taken according to the long-term plan, and only terms and the order of their flow construction were varied.

The task was to choose the least-cost flow. The constraints were the long-term plan data on the necessary growth of electric power consumption, that is, the total capacities of power plants to be launched year by year, should be no less than the total capacities stipulated by the long-term plan. It was necessary to analyze all the possible variants of the partition of the set of n power plants into flows [123].

This led to a purely combinatorial problem. For example, there are 13 variants of flows for $n = 3$. One variant provides separate construction, six variants provide construction in one flow (the order taken into account). Six more variants arise for two flows $(1 + 2)$ and their rearrangements in flows. For $n = 10$, the number of variants increases up to ten million. Let us recall that it was necessary to solve the problem for $n = 24$, which yielded more than 10^{22} variants for most primitive calculations.

It appeared impossible to fully apply the method of sequential analysis of variants to such a combinatorial problem; nevertheless, its ideas allowed developing an algorithm for eliminating nonoptimal variants and leaving approximately five million variants.

To this end, all the sites were represented by the nodes of some directed graph. Two nodes in this graph are connected by an arc if the corresponding power plants appear in one flow and are constructed successively, the orientation of the arc specifying their order. Then each variant of the decomposition into flows can be presented as a planar graph that consists of paths (oriented chains in one of two directions) of different lengths and of isolated nodes (if a flow contains only one power plant). It was proved that the graph corresponding to the optimal variant satisfied the following conditions: It has no self-intersecting paths; any two paths do not intersect; each path has the shortest length relative to its nodes (in geographical coordinates). If one of such conditions is satisfied, then the variant can always be improved; therefore, it is rejected. Based on these theoretical considerations, a general approach and relevant software were developed at the Institute of Cybernetics in 1962. This allowed finding about five best variants, from which one was chosen after discussions with construction experts and was introduced into the long-term plan. The accomplishment of this plan saved much public funds and material resources.

Applying network methods of planning and management played an important role for the national economy. First in the UkrSSR, the concept of a network model was announced at the Institute of Cybernetics at the beginning of 1963, when the Department of Economic Cybernetics received materials on nuclear submarine construction management in the US military department. In half a year, these materials became public and were studied in many scientific and research cybernetics institutes of the USSR.

The *network model* of a package of interrelated works (a project for brevity), as a rule, includes a network, that is, an *n*-node directed graph and a number of characteristics (duration, cost, resources, etc.) associated with individual works or the project as a whole. Different forms of network representation are known. Most popular is the *graphic* representation on a plane, which is called network graph. Other forms are *digital, tabular,* and those based on *light indicators, mechanical models, electronic circuits,* etc. However, all these forms are equivalent with respect to the information contained therein. In the graphic representation, the network nodes are called *events*, and the arcs correspond to *works*.

An *event* (node) means, first, a set of conditions that allow beginning one or several works (arcs) of the project that leave this node and, second, ending one or several works that enter this node. Two special events are distinguished in the network graph: *input* or *initial* event (a node with none arc entering), which has number 0, and *terminating* or *final* event (a node with none arc leaving), which has number $n - 1$. All other events are called intermediate. A terminating event is simultaneously called *target*, which means that the objectives of the whole set of works are achieved; some intermediate events can be target events.

A *work* (arc) that connects two events i and j is designated by (ij) and represents a specific working process or an expectation process. There are time, cost, resource, and some other characteristics of works. The duration of work (denoted by t_{ij}) is a time characteristics. It can only be set on the assumption that the work is performed at a constant speed. The duration can be a deterministic or a random variable, which is defined by the law of its distribution (or density function). On the graph, t_{ij} is written above the arc (ij). We will interpret the duration of work as the length of the appropriate arc.

The cost of work depends on its duration and performance conditions. The function that describes this dependence (so-called cost–time function) can be found from the analysis of different variants this work is performed or from statistical data acquaintance and processing.

Along with actual works for which resources and time are spent, the network graph may contain dummy works that mean either expectation or time dependence between works, where the beginning of one work depends on the results of another work. On the graph, dummy works have zero duration and are denoted by a dotted line.

Figure 5.3 exemplifies a network graph with ten nodes.

Given a network graph, it is possible to calculate all the necessary parameters for each event or work. Let the initial event took place at zero time. Consider an

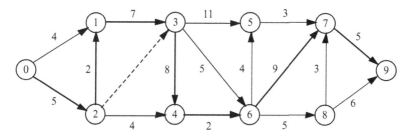

Fig. 5.3 Network model with ten events

arbitrary event i and determine the moment when this event can take place. Obviously, this moment will be no earlier than all the works from this event are completed. Following the same line of reasoning, we arrive at the conclusion that the time of event i depends on completing all the works that lie on the path from the initial event to the event i. This means that the earliest time t_i^p of the event i is equal to the length of the longest path. Calculating the early time of the final event, it is possible to calculate the minimum duration of the project (which will allow the project management to appoint a directive term) and to find out the longest path that connects the initial and final events. Obviously, at least one untimely work on this path, called critical, violates the directive time for the project completion. In Fig. 5.3, this path is denoted by bold arcs and sequentially runs through the nodes with numbers 0, 2, 1, 3, 4, 6, 7, and 9. Following the same line of reasoning, we can calculate the late completion time t_i^n for each event, that is, the deadline, which does not influence the directive project termination date. The complete design of a network graph also involves obtaining information on different time reserves for events and works. The design of large graphs is impossible without a computer. To use network graphs, it is necessary to develop an algorithm to calculate their parameters and relevant software.

An algorithm and a relevant program to design network graphs that contain no more than 3,000 events and 3,800 works were developed at the Institute of Cybernetics at the end of 1963 based on the method of sequential analysis of variants. It was the first such program in the Soviet Union. This created the basis to introduce network planning and management methods to the national economy.

Network methods and computers were applied to manage the construction of large objects such as Burshtyn GRES power plant, Lisichansk chemical plant, and the metro bridge across the Dnipro river in Kyiv at the beginning of 1964 at the Institute of Cybernetics and the Research Institute for Computer-Aided Systems in Building Industry (RICASBI) of Derzhbud of the UkrSSR under the guidance of Academician V. M. Glushkov and with the participation of V. S. Mikhalevich and V. I. Rybal'skii. Related network graphs were designed for each of them and allowed the management to amend the construction plans developed earlier, to appoint new directive terms of their accomplishment, and to optimize organizational measures. This completed the first stage of introducing the new approach – the stage of planning and preliminary management. The stage of management of the construction process involved objective difficulties.

Fig. 5.4 Event and its parameters

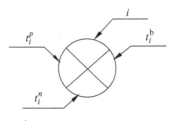

Management provides exchange of information about whether works are completed or not on time, which can lead to unforeseen changes in the network graph and necessitate the management to intervene and eventually necessitates modifying the network graph and recalculating its parameters. The latter was entrusted to a group of three programmers, employees of the Institute of Cybernetics. This information was in hard copy and transferred to punch cards (up to 300 pieces), then recalculations were performed using a computer (M-20 then) and the results were printed out. Considering human errors and possible technical faults because of the unreliability of hard copies, the whole process of recalculations lasted almost a week. Because of the specific features of construction operations, the period of information updating had not to exceed 2 weeks. Thus, the group was hard at work, and this could cause serious failures at any time.

To avoid such a situation, a manual method of recalculating network graphs without a computer was proposed. The idea was that the changes in construction that occur with a small time interval are related to a comparatively small scope of works and do not influence (almost 90 %) the parameters of the part of the network graph that does not contain these works. A backward time parameter t_i^b of realization of event i was introduced. It did not depend on the previous events but allowed obtaining the whole information on the network graph for the future in minimum time. Therefore, it was sufficient to recalculate all the changes in the curve only for the scope of works that lasted 2 weeks in this case. To implement this approach, copies of the network graphs 2 m high (to be placed on the wall) were sent to the Institute of Cybernetics, where each event with the number i was represented by a four-sector circle (Fig. 5.4) with indication of the parameters i, t_i^p, t_i^n and t_i^b. A teletype directly connected to each construction object was installed. Obtaining information from a teletypewriter tape, the members of the group covered the scope of works for two last weeks with a transparent tracing paper, recalculated the graph manually, and sent new results by teletype. Such work done for one object in no case took more than one working day, though the sizes of the network graphs were impressive: up to 2,900 events for Burshtyn GRES power plant, up to 2,400 for Lisichansk chemical plant, and up to 900 for the metro bridge. Eventually, with the advent of the paperless technology, manual calculations became unnecessary; however, the idea of the algorithm was implemented in software, and calculations were carried out on site.

The results obtained in construction showed that network planning and management methods are rather efficient to manage works and increase their productivity. Preschedule launching of some units of the Burshtyn GRES power plant saved

1.5 million rubles, and carbamide block of the Lisichansk chemical plant was constructed in 18 months instead of 30 as per the contemporary standards. Moreover, almost 70 objects were constructed in Ukraine with the participation of the Institute of Cybernetics.

The experience of the Institute of Cybernetics received deserved recognition in the Soviet Union. The manual calculation method received wide acceptance, and the four-sector representation of events even became a part of the standard instruction on drawing up network graphs issued by Derzhbud of the USSR in 1964 and has been replicated in relevant textbooks till now, though not needed for a long time. In 1964, a committee was created at the Council of Ministers of the USSR to coordinate the introduction of network planning and management methods. Mikhalevich was unanimously elected its chairman. G. A. Donets and N. Z. Shor were also elected members of the committee. Three times a year, the committee had exit sessions in the capitals of republics of the USSR, and for the whole its history (till 1969) executed a great scope of works that considerably sped up the introduction of network methods in the Soviet Union.

In 1967 in Kyiv, Mikhalevich organized the first All-Union conference on mathematical problems of network planning and management, where the achievements in this field were summarized and upcoming trends were outlined. It was mentioned at the conference that despite numerous positive examples, network methods were not successful at some construction sites. In most cases, it was because these methods did not correspond to the existing planning systems and external (mainly supplying) organizations did not participate or formally participated, attempts were made to apply network methods without understanding their essence, there was no desire and skills to abandon traditional ways of management. To eliminate these shortcomings, new, more flexible approaches to network methods were outlined, which marked the passage to automated control systems (ACS) in the industry and other branches of the national economy of the USSR.

5.2 Methods to Solve Extremum Problems on Graphs and Combinatorial Configurations

The present stage of development of the Ukrainian economy is characterized by new types and forms of economic relations among economic entities. Elaborating mechanisms to introduce them, plan, and efficiently apply is important for various branches of economy. Mathematical modeling is one of the most effective tools for the analysis and optimization of systems and processes in economy.

A graph as a mathematical concept appeared a simple, available, and powerful tool for the mathematical modeling of complex systems. Many problems for discrete objects are formulated in terms of graph theory. Such problems arise in the design and synthesis of control systems, discrete computing devices, integrated circuits, communication networks and Internet networks, in the analysis of automata, logic circuits, program flowcharts, in economics and statistics, chemistry and biology, scheduling and game theories, in pattern recognition theory and linguistics, etc. Graph theory often promotes the use of mathematical methods in various areas of science and technology, social sphere, and economics.

The development of graph theory is closely related to that of combinatorics, and the cooperation of experts that represent both fields of knowledge lasted more than one and a half century and produced the concept of combinatorial configuration. In the 1930s, classical objects of combinatorics such as incomplete counterbalanced flowcharts, tactical configurations, etc., came to be applied in design of experiments and optimal codes. The advent of computers extended the scope of application of these sciences.

In the 1970s, the set-theoretic terminology changed over to the modern graph-theoretic terminology. In 1976, R. M. Wilson finally convinced mathematicians that the majority of combinatorial problems reduced to decomposing (or covering) complete graphs into subgraphs of certain type among which so-called trees are most important.

Consider an arbitrary communication network where each of the N subscribers has direct access to the others. Some communication lines sometimes fail; therefore, the whole network should be tested periodically. One test session involves all the N points and consists in choosing a connected configuration with $N - 1$ communication lines. It is required to perform the test so as to check all the lines but none of them twice.

Analyzing the problem statement, we conclude that since the configuration of $N - 1$ lines must be connected, it can only be a tree; since the total number of lines is $\frac{N(N-1)}{2}$, the number of test sessions is $N/2$ because N is even. Thus, the problem reduces to covering $N = 2k$ isolated nodes with k N-node trees to obtain a complete N-node graph.

Let us introduce some definitions. Let there be given a complete graph H and a family of ordinary graphs $G = \{g_1, g_2, \ldots, g_k\}$. The *decomposition* of a complete graph H into graphs of family G ((H, G)-decomposition) is the partition of the set of ribs of the graph H into subgraphs (*components* of the decomposition) each being isomorphic to one of the elements of the set G. The total number of components in the decomposition is called its *dimension* (or *rank*).

The decomposition of the graph H whose components are factors of the graph H is called its *factorization*. Let us consider an example. Let $H = K_6$, and the family G consists of a single graph, a tree T of order six. This tree and the components of the corresponding (K_n, T)-decomposition are shown in Fig. 5.5. This decomposition is a factorization of the graph K_6. Factorization problems arise, for example, in motion planning for robots that operate in extreme situations in which people cannot work (fire, radiation field, explosion hazard zone, etc.). Their links constitute a complete graph, and one test session is one expedition to the hazardous area.

The factorization of a complete graph with all components isomorphic to some tree T is called *T-factorization*. A *semisymmetric* tree is a tree of order $n = 2\,k$ that has a central rib and allows an isomorphism that interchanges its ends. It is obvious

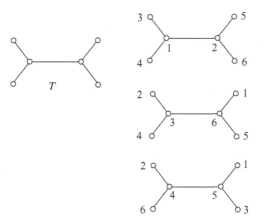

Fig. 5.5 Tree T and (K_6, T)-decomposition

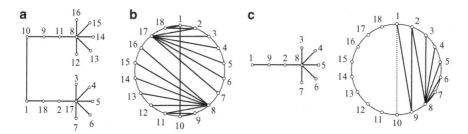

Fig. 5.6 Semisymmetric tree (**a**), correct inscription into a circle (**b**), into a semicircle (**c**)

that after the removal of the central rib, a semisymmetric tree breaks up into two isomorphic connected components (symmetric halves). They can be considered as rooted trees whose roots are the ends of the central rib of the corresponding semisymmetric tree. There is a biunique correspondence between a semisymmetric tree and its symmetric half. One of the main methods of constructing the *T-factorization* of semisymmetric trees is *the half-turn method*. The paper [10] introduces the concept of inscribing a semisymmetric tree into a circle (Fig. 5.6).

It is easy to verify that for successful inscription of a k-node tree into a semicircle, the nodes should be assigned numbers $x_i \in \{1, 2, \ldots, k\}$ so that the absolute differences of the codes of adjacent nodes form the set $\{1, \ldots, k - 1\}$. This problem was first formulated by A. Rosa and is called Rosa's problem. The problem was solved for many simple classes of trees.

The monograph [40] addresses discrete mathematical models represented as problems of edge cover of a graph or a hypergraph with typical subgraphs. These models are intended to solve important microeconomic problems such as the analysis and efficient cropland utilization in agriculture, formation of a rational portfolio of orders in leasing schemes of supply with farm machinery, print advertising, etc.

Analyzing the properties of problems and developing efficient algorithms are one of the main trends in the development of the theory of discrete extremum problems, including problems on graphs. This is because (despite, e.g., the seeming simplicity of problems on graphs) such problems have the fundamental computational complexity, and thus, it is impossible to develop exact polynomial algorithms to solve them. Even more severe difficulties are encountered in mathematical modeling of discrete systems and processes when both objectives (given many optimization criteria) and input data are uncertain.

Along with the development of efficient algorithms, it is important for such problems to analyze the structure and properties of the set of solutions, to separate out subclasses for which efficient algorithms are possibly available, to develop algorithms, to estimate their labor intensity, etc.

The problems formulated and stated in [40] were successfully analyzed by G. A. Donets and his disciples. It was proved that there exist T-factorizations for graphs with $n < 15$ nodes; Rosa's problem was solved for all trees with $n < 33$. New methods are developed to considerably expand the range of trees for which the above properties are true. Two postgraduate students who studied at the department prepared master's theses based on this subject.

Studying graphs of polyhedra involves a great number of problems that are of interest not only for graph theory, combinatorics, and topology but also for linear programming theory. Using the properties of graphs of combinatorial polyhedra may improve the efficiency of traditional methods of combinatorial optimization and promote the development of new methods. Combinatorial models can be used to represent optimization problems that arise in optimal arrangement on graphs. The combinatorial theory of polyhedra studies the extremum properties of a polyhedron and considers the set of its faces of all dimensions as some complex. But solving the above-mentioned problems encounters difficulties associated with the complexity of mathematical models, great amount of information, since the majority of problems on combinatorial sets are NP-complete. Most problems on graphs involve determining connectivity components, searching for routes, distances, etc. Important results have been currently obtained in the analysis of different classes of combinatorial models and the development of new methods of their solution. This is also true of problems on various combinatorial configurations of permutations, combinations, and arrangements [39].

Let us consider a combinatorial optimization problem: find the extremum of a linear objective function on a permutation polyhedron under additional linear constraints. As a rule, it is analyzed whether such problems can be linearized, that is, whether the convex hull of their feasible solutions can be constructed. Passing from the parametric form of the convex polyhedron to the analytic one is of great importance for discrete optimization problems since it allows formulating them in terms of linear programming; however, as practice shows, this is not always justified.

Based on [39], Fig. 5.7 shows the well-known graph of a permutation polyhedron.

Let us specify the main properties of this graph. The arrangement of points of the combinatorial set of permutations P_4 for $n \geq 4$ provides a hierarchical arrangement

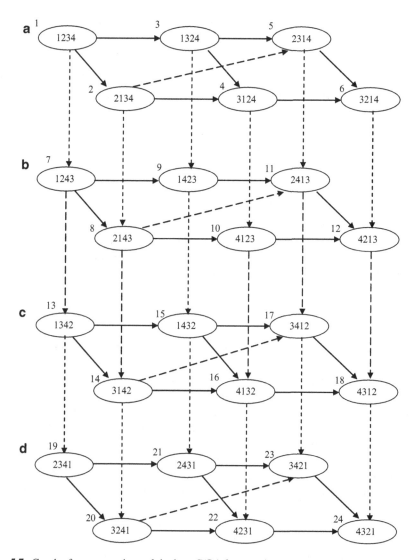

Fig. 5.7 Graph of a permutation polyhedron $G(P_4)$ for $n = 4$

of these points on the hyperplanes A, B, C, and D according to the values of the specified objective function $y^* = F(x^*)$. The graph reflects the partial order of the set of permutations for $n = 4$ with respect to the values of an arbitrary linear function $f(x) = c_1 x_1 + c_2 x_2 + c_3 x_3 + c_4 x_4$, where $c_1 \le c_2 \le c_3 \le c_4$ and the set $x = (x_1, x_2, x_3, x_4)$ runs the set of all permutations P_n.

The graph of the permutation polyhedron for an arbitrary n is constructed similarly. Two nodes of the graph that correspond to two permutations are adjacent if they can be obtained from each other by the transposition of two elements. All adjacent permutations of the graph in Fig. 5.7 are connected with arcs.

The following *statement* can be easily proved: of two adjacent permutations, the function $f(x)$ takes value in that permutation where the maximum of the two elements being permuted is on the right. This statement is true for arbitrary n.

Corollary. *The linear function $f(x)$ on the permutation polyhedron $G(P_n)$ is maximum in the permutation $(1, 2, \ldots, n)$ and is minimum in the permutation $(n, n-1, \ldots, 2, 1)$.*

The following problem often arises for such graphs: Find a set of permutations where the objective function is equal to a prescribed value, that is,

$$x^* = \arg_{x \in P_n} f(x), \tag{5.1}$$

where $f(x^*) = y$. It is also worthwhile to consider a similar problem in which permutations where the objective function takes a specified value do not always exist. Then the problem formulated above can be represented as follows: given y, determine the set of pairs of permutations $(\underline{x}, \bar{x})$ for which

$$\bar{x} = \arg \min_{f(x) > y} f(x), \quad \underline{x} = \arg \max_{f(x) < y} f(x). \tag{5.2}$$

Both problems can obviously be solved if the graph $G(P_n)$ is supplemented with a set of arcs between nonadjacent permutations, which allows passing along arcs from the initial node, where $f(x)$ is maximum, to the end node, where $f(x)$ is minimum, and visiting all the remaining nodes of the graph. This path is called *Hamiltonian*. If the sequence of permutations it passes is known, then applying the dichotomy method to the Hamiltonian path and calculating the value of the function at the corresponding permutation, it is always possible to localize an arbitrary value of the objective function $f(x)$. The complexity of the solution of problems (5.1) and (5.2) is proved to be majorized by a polynomial of degree no greater than two.

Let us show that the graph $G(P_4)$ can be constructed by induction, beginning with the first two permutations. Suppose that nodes p_1 and p_2 represent a subgraph of the graph $G(P_4)$ whose last two elements 3 and 4 are fixed. The node p_2 is formed from p_1 by transposing elements 1 and 2; therefore, $f(p_1) \geq f(p_2)$. Permuting elements 2 and 3 at p_1 and p_2 yields nodes p_3 and p_4 for which relation $f(p_3) \geq f(p_4)$ holds. Moreover, according to the above-mentioned statement, we obtain $f(p_1) \geq f(p_3)$ and $f(p_2) \geq f(p_4)$. Similarly, permuting elements 1 and 2, we obtain nodes p_5 and p_6 from p_3 and p_4. This results in a subgraph A that contains all the permutations P_4 with fixed fourth element $x_4 = 3$. Obviously, the function $f(x)$ in the subgraph A is maximum at the node p_1 and is minimum at the node p_6. Nevertheless, this subgraph has insufficient arcs to construct a Hamiltonian path. It would be possible to construct one more arc from the node p_2 to the node p_5 since these permutations differ in the transposition of numbers 1 and 3, but this is also insufficient (this arc is dotted).

Let us now take all the permutations of the subgraph A and make a simultaneous transposition of numbers 3 and 4 therein. As a result, we obtain a subgraph B that contains all the permutations P_4 with fixed fourth element $x_4 = 3$. The corresponding nodes of the subgraphs A and B are adjacent and the arcs connect them from top (from the subgraph A) to bottom (to the subgraph B). The internal orientation of the subgraph B obviously keeps the internal orientation of the subgraph A. The transposition of numbers 2 and 3 in all the permutations in the subgraph B will now yield a subgraph C that contains all the permutations P_4 with fixed fourth element $x_4 = 2$. Obviously, this subgraph also keeps the internal orientation of the subgraph B (and of the subgraph A), and an arc from the corresponding node of the subgraph B enters each of its nodes. Finally, the transposition of numbers 1 and 2 in each permutation in the subgraph C yields a subgraph D that contains all the permutations P_4 with fixed fourth element $x_4 = 1$.

All the four subgraphs A, B, C, and D compose the graph $G(P_4)$. It is said to be constructed by induction beginning with two nodes p_1 and p_2. First, these nodes were as if projected twice to form the subgraph A. Then the subgraph A was projected three times, forming the whole graph $G(P_4)$.

Let us come back to constructing a Hamiltonian path in the subgraph A. As was established, the Hamiltonian path in each subgraph is determined similarly and depends only on one parameter. The Hamiltonian path for the whole graph $G(P_4)$ is determined by two parameters. Thus, all the problems for arbitrary n can be reduced to problems on subgraphs P_4.

5.3 Problems of Constructing Discrete Images and Combinatorial Recognition

In recent years, increasing interest has been expressed by scientists and engineers in pattern recognition problems. The term "pattern recognition" denotes a wide range of activity related both to vital needs and to scientific and engineering problems. Pattern recognition always implies that a certain information processing system with input and output is available. The system can receive data from different sources such as a physical object within a certain process, experimental, meteorological, economic, or some extraneous data. The information supplied to the input of the recognition system is usually rather complex and includes signals that are often coordinate- or time-dependent functions. The initial information is relatively simple; it reduces to determining one of several classes. For example, if the input information is measurements of the coefficient of mapping of a certain surface obtained by scanning, then the initial information is the names of the symbols on this surface.

The analysis of complex images often uses formal rules that indicate the way each image is composed of elements or patterns. This results in a sequence of operations of imposing patterns on the field of vision or one pattern onto another, which somehow characterizes the image. These rules are called a grammar. The pattern recognition theory developed by Ukrainian and foreign scientists includes many such grammars.

Classification involves all processes that end by identifying some class (or membership to a class) for objects or data under consideration. Such a class may be considered as the result of recognition. In this sense, pattern recognition is a kind of classification. Most recognition problems solved until recently deal with discrete classes.

Under certain conditions, image variability and the difference between images can be determined exactly. If images can be characterized by k attributes, each having a continuous or discrete value, then it is possible to specify a k-measurable vector space (attribute space) whose each coordinate represents one attribute. In this case, an image is specified by a point in a k-measurable space.

Complex images lead to important problems that are unreasonable to be considered as classification. V. A. Kovalevskii and R. Narasimhan considered the image description problem involving automatic determination of the necessary geometric characteristics of complex images. An example is automatic processing of photographed traces of particles in bubble chambers, which is a very important problem in physical research. The photo should be analyzed to recognize the traces of particles against the noise and to determine their parameters such as direction, curvature, length, angle of propagation, etc. It is proposed to analyze characters in a similar way to describe them in certain geometrical terms, which can then be used for recognition.

The analysis of complex images is based on formal rules that indicate how images can be composed of certain elements. These rules are a kind of formal grammars and impose some constraints both on the images they produce and on the descriptions obtained by solving the problem. Thus, a formal language for image description is related to the structure of images of the class under consideration.

To imagine a two-dimensional image as a sequence, it is necessary to consider the process of forming a complex image from elementary images or elements. Elements drawn in a certain order to produce a complex image form a sequence associated with this image. Imagine standard thin transparent plates with elementary images drawn on each of them. Each terminal symbol b of the alphabet is associated with an elementary image, the parameter b characterizing this image. The grammar rules tell us which plate should be put on the previously stacked plates. Stacking all the chosen plates produces a complex image E according to the rules.

Let images be described by N-measurable vectors whose components are the dark cells of a rectangular $N = m \times n$ lattice, where m is the number of horizontal cells (rows) and n is the number of vertical cells (columns). A rectangle divided into N cells is called a field of vision π. A complex image is composed from elementary images using a componentwise operation, which may be an ordinary addition of components or their disjunction if components are binary, or any other operation. Any image on a plane can be bounded by a rectangle and then divided by horizontal and vertical straight lines into small cells, each painted in one color.

Let a set of templates $S = \{s_1, \ldots, s_p\}$ be given, where s_t $(t = 1, \ldots, p)$ is a connected figure of colored cells and $r(t)$ is the number of such figures. Each cell of the template is associated with a number that belongs to the set $Q \backslash \{0\} = \{1, \ldots, K - 1\}$. Each cell of the template is said to be colored in color $q \in Q \backslash \{0\}$. The set of

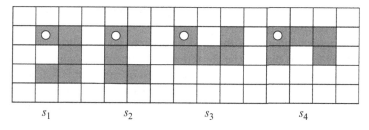

s_1 s_2 s_3 s_4

Fig. 5.8 Template and its possible positions

colors B of the field of vision π (field for short) is given, which is called the set of images (mosaics). The binary commutative operation $\oplus$ of color adding is introduced on the set Q; it associates arbitrary $a \in Q$ and $b \in Q$ with some $c \in Q$. This operation applies to each cell of the field on which templates are imposed and satisfies some conditions. More often, such an operation is modulo addition. If a template is imposed on the field, the cells will get the color that is equal to the sum of the former color and the color of the corresponding cell of the template. Several templates can be imposed on one place. The task is to use available templates to obtain any prescribed image on the uncolored field (color 0).

In each template of type s_t, one cell can be chosen that fixes the template on the field in horizontal (or vertical) position (Fig. 5.8).

Such a cell is called a template label. If the label of a template s_t is fixed λ times on the cell at the field $l = n(i - 1) + j$ $(i = 1, 2, \cdots, m, \; j = 1, 2, \cdots, n)$, denote $x_l^{(t)} = \lambda$, otherwise $x_l^{(t)} = 0$. If the template is imposed on the field vertically, then the variable $y_l^{(t)}$ will correspond to it. If the color adding operation is chosen to be addition of the numbers of colors mod K, then the problem reduces to the following mn equations:

$$\sum_{t=1}^{p} (x_l^{(t)} + y_l^{(t)}) \equiv b_l \; (\text{mod } K), \quad l = 1, 2, \ldots, N.$$

Let us formulate the problem for arbitrary addition operations.

Given sets of templates S and of colors Q and the operation of color addition, find such a subset $S' \subseteq S$ and so impose these templates on the field that the addition of colors produces a mosaic $b \in B$, where $b = \{b_1, b_2, \cdots, b_N\}$.

This problem in general form is very complex since the number of variables is much greater than the number of equations.

The theory of discrete images is well developed and has various applications, including cryptography, as a vivid example.

The use of cryptographic methods in computer-aided systems is of current importance, which was mentioned in Chap. 3. On the one hand, computer networks that transmit great amount of state, military, commercial, and private information are finding increasing use. This process should be secured against unauthorized access. On the other hand, it is necessary to take into account that new powerful supercomputers are capable of discrediting cryptographic systems, which considered almost perfect not too long ago.

Theoretical developments in the field of discrete images allow proposing new cryptographic methods that can be classed among Vigenère substitutions. In such a substitution, a key sequence of length K is imposed on the input text X, and addition modulo Z of the corresponding elements of the sequence yields a ciphered text Y at the output. There is a two-dimensional substitution where the key is an arbitrary template or even sequence of templates imposed on the text. This way can be successful even if relatively small templates are used. Note that this encryption method is intended for information that will lose its value after a rather short time after transmission, such as information transmitted by cellular phones. A small programming device for decoding and encoding that can be compactly connected to the phone can reliably protect the necessary information against unauthorized access.

5.4   Number Graph Theory

Computers are known to have been created due to the urgent need for fast computing for major scientific and technical problems in nuclear physics, aircraft engineering, climatology, etc. But their use nowadays implies absolutely different activity that involves no or little computing.

As shown in [16], each program that describes an algorithm can be associated with a directed graph. This graph is actually a *network graph,* which has already been mentioned. And vice versa, a directed acyclic graph can be considered a graph of some algorithm. It seems that well-known algorithms can be used to design network graphs. Nevertheless, the majority of the graphs of algorithms presented in [16] (see, e.g., Fig. 5.9) have a simple structure, either symmetric or consisting of identical repetitive fragments. Such graphs may have many nodes; input data of such a graph stored in a computer in a traditional way occupy much memory, which is not always convenient, hence the idea to compress information on such graphs.

In the last three decades, there have been many studies on graph and hypergraph numbering and matrix ordering. Depending on the specific field, they considered different classes of graphs and various optimality criteria.

In the general case, the majority of optimal numbering problems for graphs and matrices are NP-complete; this is why there are no efficient algorithms to find their exact solution. In this situation, it seems to be relevant to study special cases of numbering problems formulated on bounded classes of graphs. It is important both to determine polynomially solvable cases of the problem and to find the bounds of the simplifications or constraints that keep it NP-complete.

Numbering (ordering) of an n-node graph $G(X, Y)$ is a biunique mapping $f : X \Rightarrow I$, where $I = \{i_1, i_2, ..., i_n\}$ is a set of integers and f generally belongs to a class of functions K. The general numbering problem can be formulated as follows: Given a graph G, find a numbering $f \in K$ that minimizes the functional $B(G, f)$.

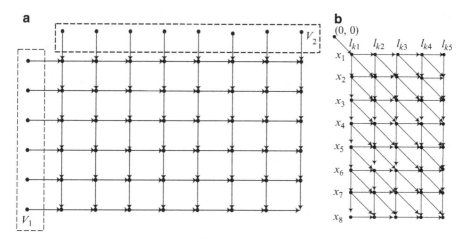

Fig. 5.9 Two-dimensional lattice graph (**a**) and typical word deployment graph (**b**)

Graph numbering theory includes a field where numbering is used to optimize the representation of graphs. The ordinary representation necessarily enumerates the nodes and ribs of a graph. There are problems where it is very difficult to represent a graph as its dimension increases; therefore, there is a need for a more compact and convenient representation.

For some discrete objects, it is important to construct the complete set of its symmetries or implement it in natural Euclidean spaces. Such problems arise in considering graphs of chemical compounds, formulas, or schemes that implement Boolean functions, digital automata, polyhedra, etc.

The problem of finding symmetries, in turn, underlies the quantitative estimate of the informativeness of discrete objects applied in scientific research related to the development of various computer-aided design systems. In particular, the complexity of an object has to be often estimated in the field of computer systems automatization, in minimizing the description of output objects, and determining informative attributes in pattern recognition.

Coding graphs to solve such problems is usually difficult; therefore, of relevance is special graph coding with the following properties: (1) It should be easy to go over from ordinary to special coding. (2) The labor input of algorithms for graphs with such coding should be lower than that with standard coding.

In view of these requirements, a coded graph should be represented as a triple $G = (X, U, F)$, where $X = \{x_1, x_2, ..., x_n\}$ is a set of real numbers, each corresponding to one node, $U = \{u_1, u_2, ..., u_m\}$ is a set of real numbers called generatrices, and F is a function of two variables such that $F(x_i, x_j) \in U$ if and only if the pair of nodes (x_i, x_j) forms a rib of the output graph. Such graphs are called *number graphs*.

Imposing different constraints on the sets X and U and function F, it is possible to obtain graph coding that reflects one property of this graph or another. In contrast to the ordinary graph representation where almost all operations are reduced to searching for an element in a data set, the basic operations in this case are reduced to evaluating the function F.

Taking $F = x_i + x_j$, $X \in N$ yields the graphs called *arithmetic* graphs or *A-graphs*. Arithmetic graphs are used in various practical problems related to the optimization of computer memory and to the improvement of computational algorithms on graphs.

Taking $F = |x_i - x_j|$, $x_j \in N$ yields so-called *modular* graphs or *M-graphs*.

Many current studies on pattern recognition, structural and systems analysis, development of software systems, design of computer facilities, and other fields employ graphs of specific structure as objects of input data. Many researchers use the ordinary graph representation, which requires huge memory and leads to cumbersome computations with large time consumption.

These graphs are mostly of hierarchical or symmetric structure consisting of repeated similar parts. Nevertheless, it is numerical graphs that allow maximally effective use of these features. It is only necessary to find the most suitable function that would determine the incidence of nodes. As a result, the graph may require several tenths or even several thousandths of the memory it occupied before, and search in huge memory can be reduced to elementary operations.

Each algorithm is known to involve three stages: (1) preliminary data processing, (2) operation of the algorithm, and (3) correction of the initial data. The first stage is intended to create the structure of input data that would optimize the algorithm. The last stage is used to output the results in a form convenient for the further use. It appears that the algorithm for number graphs can be organized so as to make the first two stages unnecessary, and most of the burden falls on the third stage.

Such algorithms were mainly used in recent studies on number graphs, where considerable attention was also paid to the existence of polynomial algorithms for number graphs. The following statement is given in [42] as a hypothesis.

Hypothesis. All the well-known NP-complete problems for number graphs have a polynomial solution algorithm.

This hypothesis is based on the assumption that the subclass of number graphs takes a special place in the general class of graphs when the information on a graph is specified in a special manner and is extremely compressed. While nonpolynomial algorithms are intended for arbitrary graphs of random structure, number graphs have much in common in their structure. For elementary graphs, we have $X = N_n$, and such graphs are called *natural*. Nevertheless, some intermediate nodes sometimes do not belong to the graph; therefore, a node exclusion function is added to the generalized definition of number graphs.

A *number* graph $G = (X, U, F, g)$ is an n-node graph represented by two sets, the set of vertices $X = N_n$ and the set of generatrices $U \in N$, and by two functions, adjacency function $F(x_i, x_j)$ and elimination function $g(x)$. A node $x_k \notin X$ if $g(x_k) = 0$, and nodes $x_i, x_j \in X$ are adjacent if $F(x_i, x_j) \in U$.

The graph theory applied in various practical fields showed that it was such specific structure most graphs had. There are a lot of monographs on pattern recognition, mathematical modeling of parallel processes, theoretical fundamentals

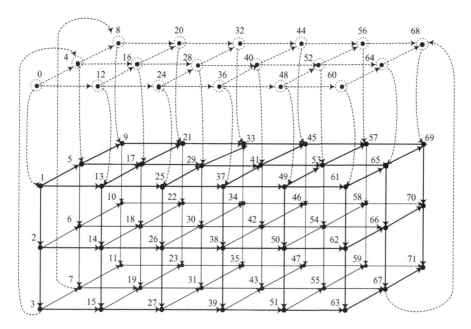

Fig. 5.10 Realization of a three-dimensional lattice graph

of the design of information technologies, etc., where graphs can be represented as number graphs in the way described above.

Let us consider, as an example, an elementary lattice graph (Fig. 5.10) used in [16]. There are algorithms whose graphs can be specified by lattice graphs of dimension three. The problem of multiplying two rectangular matrices may lead to such an algorithm as that presented in [40].

Figure 5.10 shows a graph as an M-graph, where, $n = 71, X = \{0, 1, 2, \ldots, 71\}$, $g(x) \equiv 0(\mathrm{mod}4)$, $F(x_i, x_j) = x_i - x_j$, and $U = \{1, 4, 12\}$. Dummy nodes representing the upper layer can easily be determined by the divisibility by 4. The graph is specified by only six numbers $\{71; 1; 2; 4; 0; 4\}$, whereas an array of approximately $4n$ numbers was needed earlier.

The postgraduate students trained at the Economic Cybernetics Department of the V. M. Glushkov Institute of Cybernetics defended three Ph.D. theses on this subject.

5.5 Theory of Mathematical Safes

An analysis of discrete optimization problems is a precondition for successful modeling of important economic, natural, social, and other processes. Scientific studies on discrete optimization published during the last 20–30 years are indicative of their necessity and importance. Solving problems during decision making in production process planning and management, problems of geometrical design, advanced planning, scheduling theory, and others related to choosing an alternative action has recently become of increased importance. Problems on sets located at graph nodes occupy a special place among the class of combinatorial optimization problems. Analyzing these studies, we can conclude that the new approaches and methods of combinatorial optimization are of relevance.

Let us turn to a new promising subject directly related to one of the areas of game theory. We mean the problems where, given the initial state of an object and certain rules, it is necessary to reach another prescribed state with minimum losses. Such problems arise in computer games and are generally called positional games. Mathematically, however, these are problems of mathematical safes. Let us consider a matrix B representing a safe. The elements of the matrix represent locks. An element is equal to 0 if the respective lock is open and to 1 if it is closed. After inserting a key in a lock and turning it once, all the zeroes in the same row and column become unities and vice versa. The task is to open the safe in a minimum number of such steps, that is, to zero all the elements in the matrix. The sequence of actions below does the trick. The lock with a key is underlined.

$$B = \begin{pmatrix} 1 & 0 & 0 & \underline{1} & 0 & 1 \\ 0 & 0 & 1 & 1 & 0 & 0 \\ 0 & 1 & 1 & 1 & 1 & 0 \\ 1 & 0 & 1 & 0 & 0 & 1 \end{pmatrix} \Rightarrow \begin{pmatrix} 0 & 1 & 1 & 0 & 1 & 0 \\ 0 & \underline{0} & 1 & 0 & 0 & 0 \\ 0 & 1 & 1 & 0 & 1 & 0 \\ 1 & 0 & 1 & 1 & 0 & 1 \end{pmatrix}$$

$$\Rightarrow \begin{pmatrix} 0 & 0 & 1 & 0 & 1 & 0 \\ 1 & 1 & 0 & 1 & \underline{1} & 1 \\ 0 & 0 & 1 & 0 & 1 & 0 \\ 1 & 1 & 1 & 1 & 0 & 1 \end{pmatrix} \Rightarrow \begin{pmatrix} 0 & 0 & 1 & 0 & 0 & 0 \\ 0 & 0 & 1 & 0 & 0 & 0 \\ 0 & 0 & 1 & 0 & 0 & 0 \\ 1 & 1 & \underline{1} & 1 & 1 & 1 \end{pmatrix}$$

$$\Rightarrow \begin{pmatrix} 0 & 0 & 0 & 0 & 0 & 0 \\ 0 & 0 & 0 & 0 & 0 & 0 \\ 0 & 0 & 1 & 0 & 0 & 0 \\ 1 & 0 & 1 & 1 & 0 & 0 \end{pmatrix} \Rightarrow B_{jn}.$$

Let $X = (x_{ij})_{m,n}$ be the solution, where x_{ij} is equal to the number of key turns in the lock (ij). Then the condition whereby the element b_{ij} of the matrix X becomes zero can be written as

$$\sum_{k=1}^{n} x_{ik} + \sum_{k=1}^{m} x_{kj} + b_{ij} \equiv 0 \pmod{2}, \tag{5.3}$$

where $i = 1, 2, \ldots, m$ and $j = 1, 2, \ldots, n$.

Denote by $\vec{x} = (x_{11}, x_{12}, \ldots\ldots, x_{1n}, x_{21}, x_{22}, \ldots\ldots, x_{2n}, \ldots\ldots, x_{m,n-1}, x_{mn})$ the column vector obtained from the matrix X by sequential writing of its rows. A column vector $\vec{b}$ can be obtained from the matrix B in a similar way. Moreover, let

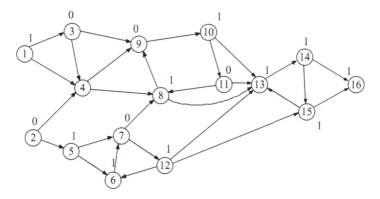

Fig. 5.11 A safe as a network

$\mathfrak{I}_n$ be an $n \times n$ matrix that consists of unities, E_n be a unit matrix of the same dimension, and I_n be a row vector of n unities. Then the transformation condition (5.3) for the whole matrix B is written as the system of equations

$$A\vec{x} \equiv \vec{b} \ (\mathrm{mod}\, 2),\tag{5.4}$$

where A is an $mn \times mn$ matrix, which consist of m^2 cells, has a standard form, and does not depend on the values of the matrix B:

$$A = \begin{pmatrix} \mathfrak{I}_n & E_n & E_n & .. & .. & E_n \\ E_n & \mathfrak{I}_n & E_n & .. & .. & E_n \\ E_n & E_n & \mathfrak{I}_n & .. & .. & E_n \\ .. & .. & .. & .. & .. & .. \\ E_n & E_n & E_n & .. & .. & \mathfrak{I}_n \end{pmatrix}.$$

Its rank and determinant depend only on the values of m and n.

If the rank of the matrix A is mn, then the solution of system (5.4) has the form

$$\vec{x} = -A^{-1}\vec{b} \ (\mathrm{mod}\, 2).$$

Similarly, a safe can be specified on a graph where the nodes are locks, and the key changes the state of the adjacent locks. Figure 5.11 exemplifies such a safe. The problem is assumed solved if all the locks are in zero state simultaneously.

To solve the safe problem, system of Eq. 5.3 should be set up for each lock. This system of equations differs from the corresponding system for matrices; therefore, it may have no solutions. We can obtain an example of a generalized safe where each lock has several states, the locks not being necessarily identical. Let us formulate the general safe problem as in [37, 38].

A mathematical safe is a system $S(Z, b, \langle Z \rangle)$ that consists of a set of locks $Z = \{z_1, z_2, \ldots, z_N\}$, initial state vector $b = (b_1, b_2, \ldots, b_N)$ (where $b_i \in \{0, 1, \ldots, k_i - 1\}$ is the state of the ith lock), and the set $\langle Z \rangle = \{Z_1, Z_2, \ldots, Z_N\}$,

$z_l \in Z_l$, $Z_l \in 2^Z$ ($1 \leq i, \ l \leq N$). One clockwise turn of a key in the lock z_l changes the states of all the locks $z_j \in Z_l$ from b_j to $(b_j + 1) \pmod{k_j}$. The safe is open if it is in the state $b = (0, \ 0, \ldots, \ 0)$. For each lock z_i, it is necessary to find the number of key turns x_i needed to open the safe.

The vector $\vec{x} = (x_1, \ x_2, \ldots, \ x_N)$ is called the solution of the safe problem. The set $\langle \mathbf{Z} \rangle$ is called the incidence set. It can be written as an $N \times N$ incidence matrix $A_0 = [a_{ij}^0]$, with zeroes on the principal diagonal, and $a_{ij}^0 = 1$ if z_j belongs to the set Z_i ($1 \leq i, j \leq N$) and $a_{ij}^0 = 0$ otherwise. The matrix A_0 can be associated with a directed graph $G(Z)$ where an arc comes from the node z_i to the node z_j if $a_{ij}^0 = 1$. Different problems arise depending on the complexity of this matrix. Denote $A = A_0 + E_N$, where E_N is a unit matrix. The column of this matrix that corresponds to the jth lock contains unities opposite the locks that influence the state of the jth lock. Considering the number of all key turns in these locks and the number x_j of key turns in the given lock, we obtain the total number of key turns in the jth lock. Summed with the initial state of the jth lock, this should be equal to 0 (mod k_j). Then the general safe problem is reduced to the system of linear equations

$$\vec{x} \, \vec{a}_i + b_i \equiv 0 \pmod{k_i}, \quad 1 \leq i \leq N,$$

where $\vec{a}_i$ is the ith column of the matrix A.

If $k_i = K = \text{const}$ for all $1 \leq i \leq N$, such locks are of one type. If safes have different types of locks, they can be described by the vector $K = \{k_1, k_2, \ldots, k_N\}$, where some elements may coincide. Denote $b(\text{mod } K) = (b_1(\text{mod } k_1), b_2(\text{mod } k_2), \ldots, b_N(\text{mod } k_N))$. The general safe problem on graphs is reduced to the system of linear equations

$$A\vec{x} \equiv -b \pmod{K}.$$

If there exists the inverse matrix A^{-1}, then $\vec{x} \equiv -A^{-1}b \pmod{K}$.

Safe problems on matrices are solved completely and for many graphs, mostly of simple structures. In the last decade, studies on this subject have actively been carried out, and Ph.D. theses have been defended.

Afterword

Telling about a man of science means first to show the results of his creative activity, which is what we were trying to do. This is because work is largely the meaning of a scientist's life. It takes the most time and effort; it also brings the most pleasure.

When the earthly path of a scientist ends, he continues to be present in public life, his ideas are working and, if lucky, developed and taken up by disciples and followers who go further. Mikhalevich was lucky with his ideas.

The creative activity of Volodymyr Mikhalevich was mainly focused on the development of various approaches to solving complex optimization problems that arise in all spheres of human activity. This was outlined in our book. In conclusion, it is important to emphasize that the development and theoretical foundation of new optimization methods is a contribution made by V.S. Mikhalevich and his disciples to modern mathematics and informatics. These methods, as well as the approaches proposed for mathematical modeling of complex processes that characterize modern life, including modern science, form the foundation for modern computer technologies, which are efficiently implemented usually on supercomputer systems.

It is very typical for works in this field that the society's interest in them is not diminished over time but is growing. That is why we are witnessing the rapid and qualitative expansion of the Ukrainian school of optimization and enhanced interest of foreign specialists in the scientific results of representatives of this school. As problems of transcomputational complexity became increasingly important, the need arose both to further improve the available optimization methods and to develop new ones. This is a natural process, and it will always be associated with the name of V.S. Mikhalevich, who has made an invaluable contribution to its development.

As the development of the world's science shows, the fields in which practice and life are directly interested develop especially rapidly. Modern informatics is among them, and computer mathematics is its core. Mikhalevich was one of the first to understand that; his pioneering works in this field arouse great interest. Serving science was the main driving force of his unfortunately short earthly life. No doubt that V.S. Mikhalevich will remain in science forever as one of its leaders.

I.V. Sergienko, *Methods of Optimization and Systems Analysis for Problems of Transcomputational Complexity*, Springer Optimization and Its Applications 72, DOI 10.1007/978-1-4614-4211-0, © Springer Science+Business Media New York 2012

References

1. Alekseev, G.V., Komarov, E.G.: Numerical analysis of extremum problems in the theory of sound radiation in a flat waveguide. Matem. Modelirovanie **3**(12), 52–64 (1991)
2. Alekseev, A.A., Alekseev, D.A.: Practical Macroeconomic Models [in Ukrainian], p. 268. Naukova Dumka, Kyiv (2006)
3. Artur, V., Ermol'ev, Y.M., Kaniovskii, Y.M.: A generalized urn problem and its applications. Cybernetics **19**(1), 61–70 (1983)
4. Babich, M.D.: An approximation-iteration method for solving nonlinear operator equations. Cybern. Syst. Anal. **27**(1), 21–28 (1991)
5. Babich, M.D., Zadiraka, V.K., Sergienko, I.V.: A computing experiment in the problem of optimization of computations. Cybern. Syst. Anal. **35**(1), 47–55, Pt. 1; 2, Pt. 2, 221–239 (1999)
6. Bakaev, O.O., Branovyts'ka, S.V., Mikhalevich, V.S., Shor, N.Z.: Determining the characteristics of a transport network by the method of sequential analysis of variants. Dop AN URSR **44**, 472–474 (1962)
7. Belan, E.P., Mikhalevich, M.V., Sergienko, I.V.: Models of two-sided monopolistic competition on the labor market. Cybern. Syst. Anal. **41**(2), 175–182 (2005)
8. Belan, E.P., Mikhalevich, M.V., Sergienko, I.V.: Cyclicity and well-balanced growth in systems with imperfect competition in labor markets. Cybern. Syst. Anal. **43**(5), 29–47 (2007)
9. Belan, E.P., Mikhalevich, M.V., Sergienko, I.V.: Cyclic economic processes in systems with monopsonic labor markets. Cybern. Syst. Anal. **39**(4), 24–39 (2003)
10. Belan, E.P., Mikhalevich, M.V., Sergienko, I.V.: Cycles in economic systems with open labor markets. Cybern. Syst. Anal. **44**(4), 48–72 (2008)
11. Bellman, R.: Dynamic Programming. Princeton University Press, Princeton (1957)
12. Bidyuk, P.I., Baklan, I.V., Savenkov, O.I.: Time Series: Modeling and Predicting [in Ukrainian]. EKMO, Kyiv (2003)
13. Wald, A.: Sequential Analysis. Wiley, New York (1947)
14. Vasil'ev, P.F.: Solution Methods for Extremum Problems [in Russian]. Nauka, Moscow (1981)
15. Vdovichenko, I.A., Trubin, V.A., Shor, N.Z., Yun, G.N.: Problem of typification of technological objects. Cybernetics **10**(6), 1032–1034 (1974)
16. Voevodin, V.V.: Mathematical Models and Methods in Parallel Processes [in Russian]. Nauka, Moscow (1986)
17. Mikhalevich, V.S. (ed.): Computational Methods for Choosing Optimal Design Solutions [in Russian]. Naukova Dumka, Kyiv (1977)
18. Gaivoronskii, A.A.: Nonstationary stochastic programming problems. Cybernetics **14**(5), 575–578 (1978)

19. Gershovich, V.I., Shor, N.Z.: Ellipsoid method, its generalizations and applications. Kibernetika **5**, 61–69 (1982)
20. Gladkii, A.V., Sergienko, I.V., Skopetskii, V.V.: Numerical-Analytic Methods to Study Wave Processes [in Russian]. Naukova Dumka, Kyiv (2001)
21. Glushkov, V.M.: Systemwise optimization. Cybernetics **16**(5), 731–732 (1980)
22. Glushkov, V.M., Ivanov, V.V., Mikhalevich, V.S., Sergienko, I.V., Stognii, A.A.: Reserves of Computation Optimization. Preprint IC AS USSR: 77–67 [in Russian], Kyiv (1977)
23. Glushkov, V.M., Tseitlin, G.E., Yushchenko, E.L.: Symbolic Multiprocessing Methods [in Russian]. Naukova Dumka, Kyiv (1980)
24. Gnedenko, B.V., Mikhalevich, V.S.: On the distribution of the number of outputs of one empirical distribution function over another. Dokl. AN SSSR **82**(6), 841–843 (1951)
25. Gnedenko, B.V., Mikhalevich, V.S.: Two theorems on the behavior of empirical distribution functions. Dokl. AN SSSR **85**(1), 25–27 (1952)
26. Gulyanitskii, L.F., Sergienko, I.V.: Metaheuristic downhill simplex method in combinatorial optimization. Cybern. Syst. Anal. **43**(6), 822–829 (2007)
27. Gupal, A.M.: Stochastic Methods to Solve Nonsmooth Extremum Problems [in Russian]. Naukova Dumka, Kyiv (1979)
28. Gupal, A.M., Sergienko, I.V.: Optimal Pattern Recognition Procedures [in Russian]. Naukova Dumka, Kyiv (2008)
29. Daduna, H., Knopov, P.S., Tur, L.P.: Optimal strategies for an inventory system with cost functions of general form. Cybern. Syst. Anal. **35**(4), 602–618 (1999)
30. Deineka, V.S.: Optimal Control of Elliptic Multicomponent Distributed Systems [in Russian]. Naukova Dumka, Kyiv (2005)
31. Deineka, V.S., Sergienko, I.V.: Analysis of Multicomponent Distributed Systems and Optimal Control [in Russian]. Naukova Dumka, Kyiv (2007)
32. Deineka, V.S., Sergienko, I.V.: Models and Methods to Solve Problems in Inhomogeneous Media [in Russian]. Naukova Dumka, Kyiv (2001)
33. Deineka, V.S., Sergienko, I.V., Skopetskii, V.V.: Mathematical Models and Methods to Design Problems with Discontinuous Solutions [in Russian]. Naukova Dumka, Kyiv (1995)
34. Dem'yanov, V.F., Vasil'ev, L.V.: Nondifferential Optimization [in Russian]. Nauka, Moscow (1981)
35. Dikin, I.I., Zorkal'tsev, V.I.: Iterative Solution of Mathematical Programming Problems [in Russian]. Nauka, Novosibirsk (1980)
36. Dolinskii, A.A.: Energy saving and ecological problems in power engineering. Visnyk NANU **2**, 24–32 (2006)
37. Donets, G.A.: Solution of the safe problem on (0,1)-matrices. Cybern. Syst. Anal. **38**(1), 83–88 (2002)
38. Donets, G.A.: Solution of a matrix problem on a mathematical safe for locks of the same kind. Cybern. Syst. Anal. **41**(2), 289–301 (2005)
39. Donets, G.A., Kolechkina, L.N.: Method of ordering the values of a linear function on a set of permutations. Cybern. Syst. Anal. **45**(2), 204–213 (2009)
40. Donets, G.A., Ya, A.: Petrenyuk, Extremal Graph Covers [in Russian]. Combinatorni Konfiguratsii, Kirovograd (2009)
41. Donets, G.A., Samer Alshalame, I.M.: Solving the linear-mosaic problem, Teoriya Optimal'nykh Reshenii. V. M. Glushkov Institute of Cybernetics NAS Ukraine. **4**, 15–24 (2005)
42. Donets, G.A., Shulinok, I.E.: "Estimating the complexity of algorithms for natural modular graphs," Teoriya Optimal'nykh Reshenii. V. M. Glushkov Institute of Cybernetics NAS Ukraine. **4**, 61–68 (2001)
43. Emelichev, V.A., Komlik, V.I.: Method of Constructing a Sequence of Plans to Solve Discrete Optimization Problems [in Russian]. Nauka, Moscow (1981)
44. Energy Strategy of Ukraine for the period till 2030. www.mpenergy.gov.ua
45. Eremin, I.I.: Penalty method in convex programming. Kibernetika **4**, 63–67 (1967)

46. Eremin, I.I., Astaf'ev, N.N.: An Introduction to Linear and Convex Programming Theory [in Russian]. Nauka, Moscow (1976)
47. Ermol'ev, Y.M., Verchenko, P.I.: A linearization method in limiting extremal problems. Cybernetics 12(2), 240–245 (1976)
48. Ermol'ev, Y.M., Gaivoronskii, A.A.: Stochastic method for solving minimax problems. Cybernetics 19(4), 550–558 (1983)
49. Ermol'ev, Y.M., Mirzoakhmedov, F.: Direct methods of stochastic programming in inventory control problems. Cybernetics 12(6), 887–894 (1976)
50. Ermol'ev, Y.M., Mikhalevich, V.S., Chepurnoi, N.D.: Design of parallel optimization methods. Cybernetics 23(5), 571–580 (1987)
51. Ermol'ev, Y.M.: Method of generalized stochastic gradients and stochastic quasi–Fejér sequences. Kibernetika 2, 73–83 (1969)
52. Ermol'ev, Y.M.: Methods of solving nonlinear extremum problems. Kibernetika 4, 1–17 (1966)
53. Ermol'ev, Y.M.: Stochastic Programming Methods [in Russian]. Nauka, Moscow (1976)
54. Ermol'ev, Y.M., Yashkir, O.V., Yashkir, Y.M.: Nondifferentiable and Stochastic Optimization Methods in Physical Research [in Russian]. Naukova Dumka, Kyiv (1995)
55. Ermol'ev, Y.M., Gulenko, V.P., Tsarenko, T.I.: Finite-Difference Method in Optimal Control Problems [in Russian]. Naukova Dumka, Kyiv (1978)
56. Ermol'ev, Y.M., Ermol'eva, T.Y., McDonald, G., Norkin, V.I.: Problems of insurance of catastrophic risks. Cybern. Syst. Anal. 37(2), 220–234 (2001)
57. Ermol'ev, Y.M., Knopov, P.S.: Method of empirical means in stochastic programming problems. Cybern. Syst. Anal. 42(6), 773–785 (2006)
58. Ermol'ev, Y.M., Mar'yanovich, T.P.: Optimization and modeling. Probl. Kibernetiki. 27, 33–38 (1973)
59. Ermol'ev, Y.M., Nekrylova, Z.V.: Stochastic subgradient method and its applications. Teoriya Optim. Resh. Institute of Cybernetics AS USSR. 1, 24–47 (1967)
60. Ermol'ev, Y.M., Norkin, V.I.: Stochastic generalized gradient method for the solution of nonconvex nonsmooth problems of stochastic optimization. Cybern. Syst. Anal. 34(2), 196–215 (1998)
61. Ermol'ev, Y.M., Norkin, V.I.: Nonstationary law of large numbers for dependent random variables and its application in stochastic optimization. Cybern. Syst. Anal. 34(4), 553–563 (1998)
62. Ermol'ev, Y.M., Shor, N.Z.: Minimization of nondifferentiable functions. Kibernetika 1, 101–102 (1967)
63. Zhukovskii, V.I., Chikrii, A.A.: Linear-Quadratic Differential Games [in Russian]. Naukova Dumka, Kyiv (1994)
64. Zavadskii, V.Yu.: Modeling Wave Processes [in Russian]. Nauka, Moscow (1991)
65. Zadiraka, V.K., Mikhalevich, V.S., Sergienko, I.V.: Software development to solve statistical processing problems [in Russian]. Preprint AS USSR: 79–50, Kyiv (1979)
66. Zadiraka, V.K., Oleksyuk, O.S.: Computer Cryptology [in Ukrainian]. Zbruch, Kyiv-Ternopilɜ (2002)
67. Zgurovsky, M.Z.: Interrelation between Kondratieff cycles and global systemic conflicts. Cybern. Syst. Anal. 45(5), 742–749 (2009)
68. Zgurovskii, M.Z., Bidyuk, P.I., Terent'ev, A.N.: Methods of constructing Bayesian networks based on scoring functions. Cybern. Syst. Anal. 44(2), 219–225 (2008)
69. Zgurovskii, M.Z., Gvishiani, A.D.: Global Modeling of Sustainable Development Processes [in Russian]. NTUU "KPI", Kyiv (2008)
70. Zgurovskii, M.Z., Pankratova, N.D.: Systems Analysis: Problems, Methodology, Applications [in Russian]. Naukova Dumka, Kyiv (2005)
71. Zgurovskii, M.Z., Pankratova, N.D.: Technological Prediction, [in Russian]. Politekhnika, Kyiv (2005)

72. Zgurovskii, M.Z., Pankratova, N.D.: Fundamentals of Systems Analysis [in Ukrainian]. BHV, Kyiv (2007)
73. Ivanenko, V.I., Mel'nik, V.S.: Variational Methods in Control Problems for Distributed-Parameter Objects [in Russian]. Naukova Dumka, Kyiv (1988)
74. Ivakhnenko, A.G., Stepashko, V.S.: Noise Immunity of Modeling [in Russian]. Naukova Dumka, Kyiv (1985)
75. Kaniovskii, Yu.M., Knopov, P.S., Nekrylova, Z.V.: Limiting Theorems for Stochastic Programming Processes [in Russian]. Naukova Dumka, Kyiv (1980)
76. Kirilyuk, V.S.:Optimal decisions under risk based on multivalued mappings. Thesis of Dr. Phys.-Math. Sci. V. M. Glushkov Institute of Cybernetics, NAS of Ukraine, Kyiv (2006)
77. Kiseleva, E.M., Shor, N.Z.: Continuous Optimal Set Partition Problems: Theory, Algorithms, and Applications [in Russian]. Naukova Dumka, Kyiv (2005)
78. Knopov, P.S.: Optimal Estimates for Stochastic System Parameters [in Russian]. Naukova Dumka, Kyiv (1991)
79. Knopov, P.S., Kasitskaya, E.I.: Large deviations of empirical estimates in stochastic programming problems. Cybern. Syst. Anal. 40(4), 52–60 (2004)
80. Kovalenko, I.N.: Analysis of Rare Events in Estimating System Efficiency and Reliability [in Russian]. Sov Radio, Moscow (1980)
81. Kovalenko, I.N., Kuznetsov, N.Yu.: Methods to Design Highly Reliable Systems [in Russian]. Radio i Svyaz', Moscow (1988)
82. Krivonos, Yu.G., Matichin, I.I., Chikrii, A.A.: Dynamic Games with Discontinuous Trajectories [in Russian]. Naukova Dumka, Kyiv (2005)
83. Kuntsevich, V.M.: Pulse Self-Adapting and Extremal Automatic Control Systems [in Russian]. Tekhnika, Kyiv (1966)
84. Kuntsevich, V.M.: Control under Uncertainty: Guaranteed Results in Control and Identification Problems [in Russian]. Naukova Dumka, Kyiv (2006)
85. Kuntsevich, V.M., Lychak, M.M.: Synthesis of Optimal and Adaptive Control Systems: A Game Approach [in Russian]. Naukova Dumka, Kyiv (1985)
86. Kuntsevich, V.M., Lychak, M.M.: Synthesis of Automatic Control Systems Using Lyapunov Functions [in Russian]. Nauka, Moscow (1977)
87. Kuntsevich, V.M., Chekhovoi, Yu.N.: Nonlinear Control Systems with Pulse-Frequency and Pulse-Width Modulation [in Russian]. Tekhnika, Kyiv (1970)
88. Kukhtenko, A.I.: Invariance Problems in Automatics [in Russian]. Gostekhizdat УССР, Kyiv (1963)
89. Laptin, Yu.P., Zhurbenko, N.G., Levin, M.M., Volkovitskaya, P.I.: Using optimization means in the KROKUS system of computer-aided design of power steam generating units. Energetika i Elektrifikatsiya 7, 41–51 (2003)
90. Lebedeva, T.T., Sergienko, T.I.: Different types of stability of vector integer optimization problem: General approach. Cybern. Syst. Anal. 44(3), 429–433 (2008)
91. Mel'nik, V.S., Zgurovskii, M.Z.: Nonlinear Analysis and Control of Infinite-Dimensional Systems [in Russian]. Naukova Dumka, Kyiv (1999)
92. Mel'nik, V.S., Zgurovskii, M.Z., Novikov, A.N.: Applied Methods of the Analysis and Control of Nonlinear Processes and Fields [in Russian]. Naukova Dumka, Kyiv (2004)
93. Mikhalevich, V.S., Kapitonova, Yu.V., Letichevskii, A.A.: Macropipelining of computations. Cybernetics 22(3), 3–10 (1986)
94. Mikhalevich, V.S., Kapitonova, Yu.V., Letichevskii, A.A., Molchanov, I.N., Pogrebinskii, S.B.: Organization of computations in multiprocessor computers. Kibernetika 3, 1–10 (1984)
95. Mikhalevich, V.S., Kuksa, A.I.: Sequential Optimization Methods in Discrete Network Problems of Optimal Resource Allocation [in Russian]. Nauka, Moscow (1983)
96. Mikhalevich, V.S., Sergienko, I.V., Zadiraka, V.K., Babich, M.D.: Optimization of computations. Cybern. Syst. Anal. 30(2), 213–235 (1994)
97. Mikhalevich, V.S., Kanygin, Y.M., Gritsenko, V.I.: Informatics: General Principles [in Russian]. Prepr. V. M. Glushkov Inst. of Cybernetics AS USSR, 83–31, Kyiv (1983)

98. Mikhalevich, V.S.: Bayesian solutions and optimal acceptance sampling methods. Ukr. Mat. Zh. 7(4), 454–459 (1956)

99. Mikhalevich, V.S.: Bayesian choice between two hypotheses on the mean value of a normal process. Visn. Kyiv. Univ. 1(1), 101–104 (1958)

100. Mikhalevich, V.S.: On the mutual position of two empirical distribution functions. Dokl. AN SSSR 3, 485–488 (1952)

101. Mikhalevich, V.S.: Sequential optimization algorithms and their application. A sequential search algorithm. Kibernetika 1, 45–55 (1965)

102. Mikhalevich, V.S.: Sequential optimization algorithms and their application. Kibernetika 2, 85–89 (1965)

103. Mikhalevich, V.S.: Sequential Bayesian solutions and optimal acceptance sampling methods. Teoriya. Veroyatn. Primen. 1(4), 395–421 (1956)

104. Mikhalevich, V.S.: Sequential Bayesian solutions and optimal acceptance sampling methods. Abstracts of Ph.D. thesis [in Russian]. Moscow (1956)

105. Mikhalevich, V.S., Volkovich, V.L.: Computational Methods for the Analysis and Design of Complex Systems [in Russian]. Nauka, Moscow (1982)

106. Mikhalevich, V.S., Gupal, A.M., Norkin, V.I.: Nonconvex Optimization Methods [in Russian]. Nauka, Moscow (1987)

107. Mikhalevich, V.S., Kanygin, Y.M., Gritsenko, V.I.: Informatics: a new field of science and practice. In: Cybernetics. Evolution of Informatics (A Collection of Scientific Works) [in Russian], pp. 31–45. Nauka, Moscow (1986)

108. Mikhalevich, V.S., Knopov, P.S., Golodnikov, A.N.: Mathematical models and methods of risk assessment in ecologically hazardous industries. Cybern. Syst. Anal. 30(2), 259–273 (1994)

109. Mikhalevich, V.S., Kuksa, A.I.: Sequential Optimization Methods in Discrete Network Problems of Optimal Resource Allocation [in Russian]. Nauka, Moscow (1983)

110. Mikhalevich, V.S., Kuntsevich, V.M.: An Approach to Study Armament Control Processes [in Russian], vol. M. Glushkov Inst. of Cybernetics, AS USSR, Kyiv (1989)

111. Mikhalevich, V.S., Mikhalevich, M.V., Podolev, I.V.: Modeling Individual Mechanisms of the State Influence on Prices in Transition Economy [in Russian]. Prepr. V. M. Glushkov Inst. of Cybernetics, NAS of Ukraine, Kyiv (1994)

112. Mikhalevich, V.S., Sergienko, I.V., Zadiraka, V.K., et al.: Optimization of Computations [in Ukrainian]. Preprint V. M. Glushkov Inst. of Cybernetics, NAS of Ukraine; 99–13, Kyiv (1993)

113. Mikhalevich, V.S., Sergienko, I.V., Trubin, V.A., Shor, N.Z., et al.: Program package for solving large-scale production and transportation planning problems (PLANNER). Cybernetics 19(3), 362–381 (1983)

114. Mikhalevich, V.S., Sergienko, I.V., Shor, N.Z.: Investigation of optimization methods and their applications. Cybernetics 17(4), 522–547 (1981)

115. Mikhalevich, V.S., Sergienko, I.V., Shor, N.Z., et al.: The software package DISPRO-3: Objectives, classes of problems, systemic and algorithmic support. Cybernetics 21(1), 68–85 (1985)

116. Mikhalevich, V.S., Trubin, V.A., Shor, N.Z.: Optimization Problems of Production-Transportation Planning [in Russian]. Nauka, Moscow (1986)

117. Mikhalevich, V.S., Shkurba, V.V.: Sequential optimization algorithms in work ordering problems. Kibernetika 2, 34–40 (1966)

118. Mikhalevich, V.S., Shor, N.Z.: Mathematical fundamentals of the solution of problems of the choice of optimal outline of longitudinal profile. In: pp. 22–28. Trans. VNII Transp. Stroit., Moscow (1964)

119. Mikhalevich, V.S., Shor, N.Z.: Sequential analysis of variants in solving variational control, planning, and design problems. In: Abstracts of Papers Read at the 4th All-Union Math. Congress, Leningrad (1961), p. 91

120. Mikhalevich, V.S., Shor, N.Z.: Numerical solutions of multiple-choice problems by the method of sequential analysis of variants. In: Proceedings of the Econ.-Math. Seminar, Issue 1, pp. 15–42. Rotaprint AS SSSR, LEMI, Moscow (1962)

121. Mikhalevich, V.S., Shor N.Z., Bidulina, L.M.: Mathematical methods to choose the optimal complex gas-main pipeline with stationary gas flow. In: Economic Cybernetics and Operations Research [in Russian]. pp. 57–59. V. M. Glushkov Inst. of Cybernetics, AS USSR, Kyiv (1967)

122. Mikhalevich, V.S., Yun, G.N.: Optimization of the main design parameters of passenger aircraft. Visn AN USSR **8**, 39–43 (1981)

123. Mikhalevich, V.S., Rybal'skii, V.I.: Computer Design Principles for Optimal Terms and Priority in Construction [in Russian]. Akad. Stroit. Arkhitekt. SSSR, Kyiv (1962)

124. Mikhalevich, V.S., Sergienko, I.V.: Modeling of Transition Economy: Models, Methods, Information Technologies [in Russian]. Naukova Dumka, Kyiv (2005)

125. Mikhalevich, V.S., Chizhevskaya, A.Yu.: Dynamic macromodels of unstable processes in transition to a market economy. Cybern. Syst. Anal. **29**(4), 538–545 (1993)

126. Moiseev, N.N.: Numerical Methods in Optimal Systems Theory [in Russian]. Nauka, Moscow (1971)

127. Mikhalevich, V.S., Skurikhin, V.I., Kanygin, Y.M., Gritsenko, V.I.: Some Approaches to Developing the Concept of Society Informatization [in Russian]. Prepr. V. M. Glushkov Inst. of Cybernetics, AS USSR; 88–66, Kyiv (1988)

128. Nemirovsky, A.S., Yudin, D.B.: Problem Complexity and Optimization Method Efficiency [in Russian]. Nauka, Moscow (1979)

129. Norkin, V.I., Onishchenko, B.O.: Reliability optimization of a complex system by the stochastic branch and bound method. Cybern. Syst. Anal. **44**(3), 418–428 (2008)

130. Norkin, V.I., Keyzer, M.A.: Efficiency of classification methods based on empirical risk minimization. Cybern. Syst. Anal. **45**(5), 750–761 (2009)

131. Nurminskii, E.A.: Numerical Methods to Solve Stochastic Minimax Problems [in Russian]. Naukova Dumka, Kyiv (1979)

132. Ostapenko, V.V., Skopetskii, V.V., Finin, G.S.: Space and Time Resource Allocation [in Ukrainian]. Naukova Dumka, Kyiv (2003)

133. Pankratova, N.D.: Systems analysis in the dynamics of diagnosing complex engineering systems. Syst. Doslidzh. Inform. Tekhnologii **1**, 33–49 (2008)

134. Parasyuk, I.N., Sergienko, I.V.: Software Packages for Data Analysis. Development Technology [in Russian]. Finansy i Statistika, Moscow (1988)

135. Shkurba, V.V., Boldyreva, V.V., V'yun, A.F., et al.: Planning Discrete Manufacturing with ACSs [in Russian]. Tekhnika, Kyiv (1975)

136. Podchasova, T.P., Lagoda, A.P., Rudnitskii, V.F.: Control in Hierarchical Industrial Structures [in Russian]. Naukova Dumka, Kyiv (1989)

137. Polinkevich, K.B., Onopchuk, Yu.N.: Conflicts in the regulation of the main function of the respiratory system of the organism and mathematical models of conflict resolution. Cybernetics **22**(3), 385–390 (1986)

138. Polyak, B.T.: Minimization of nonsmooth functionals. Vych. Mat. Mat. Fiz. **1**(3), 509–521 (1969)

139. Mikhalevich, V.S., Volkovich, V.L., Voloshin, A.F., Mashchenko, S.O.: Sequential approach to the solution of mixed linear problems. Cybernetics **19**(1), 34–40 (1991)

140. Glushkov, V.M., Mikhalevich, V.S., Sibirko, A.N., et al.: Applying electronic digital computers to design railways. In: Tr. TsNIIS and IK AN USSR, Issue 51 [in Russian], Moscow (1964)

141. Pshenichnyi, B.N.: Convex Analysis and Extremum Problems [in Russian]. Nauka, Moscow (1980)

142. Pshenichnyi, B.N.: Linearization Methods [in Russian]. Nauka, Moscow (1983)

143. Pshenichnyi, B.N.: Necessary Extremum Conditions [in Russian]. Nauka, Moscow (1969)

144. Pshenichnyi, B.N.: Necessary Extremum Conditions [in Russian], 2nd edn. Nauka, Moscow (1982)
145. Pshenichnyi, B.N., Danilin, Yu.M.: Numerical Methods in Extremum Problems [in Russian]. Nauka, Moscow (1975)
146. Pshenichnyi, B.N., Ostapenko, V.V.: Differential Games [in Russian]. Naukova Dumka, Kyiv (1992)
147. Resolution of the National Security and Defense Council of May 30, 2008. http://www.rainbow.gov.ua
148. Rockafellar, R.: Convex Analysis. Princeton University Press, Princeton (1970)
149. Samarskii, A.A., Vabishchevich, P.N.: Computational Heat Transfer [in Russian]. Editorial URSS, Moscow (2003)
150. Samoilenko, Yu.I., Butkovskii, A.N.: Control of Quantum-Mechanical Objects [in Russian]. Nauka, Moscow (1984)
151. Sergienko, I.V.: An abstract problem formulation revisited. Dop. AN URSR **2**, 177–179 (1965)
152. Sergienko, I.V.: A method of searching for extremum solutions in one class of problems. Dop. AN URSR **3**, 296–299 (1965)
153. Sergienko, I.V.: Mathematical Models and Methods to Solve Discrete Optimization Problems [in Russian]. Naukova Dumka, Kyiv (1988)
154. Sergienko, I.V.: "A method to solve a special problem of scheduling theory", in: Algorithmic Languages and Automation of Programming [in Russian], pp. 54–62. Naukova Dumka, Kyiv (1965)
155. Sergienko, I.V., Deineka, V.S.: Systems Analysis of Multicomponent Distributed Systems [in Russian]. Naukova Dumka, Kyiv (2009)
156. Sergienko, I.V., Kaspshitskaya, N.F.: Models and Methods for Computer Solution of Combinatorial Optimization Problems [in Russian]. Naukova Dumka, Kyiv (1981)
157. Sergienko, I.V., Kozeratskaya, L.N., Lebedeva, T.T.: Stability and Parametric Analyses of Discrete Optimization Problems [in Russian]. Naukova Dumka, Kyiv (1995)
158. Sergienko, I.V., Lebedeva, T.T., Roshchin, V.A.: Approximate Methods to Solve Discrete Optimization Problems [in Russian]. Naukova Dumka, Kyiv (1980)
159. Sergienko, I.V., Shilo, V.P.: Discrete Optimization Problems: Challenges, Solution Methods, Analysis [in Russian]. Naukova Dumka, Kyiv (2003)
160. Sergienko, I.V., Shilo, V.P.: Problems of discrete optimization: Challenges and main approaches to solve them. Cybern. Syst. Anal. **42**(4), 465–483 (2006)
161. Sergienko, I.V., Yanenko, V.M., Atoev, K.L.: Conceptual framework for managing the risk of ecological, technogenic, and sociogenic disasters. Cybern. Syst. Anal. **33**(2), 203–219 (1997)
162. Sergienko, I., Koval, V.: SCIT: A Ukrainian supercomputer project. Visnyk NANU **8**, 3–13 (2005)
163. Sergienko, I.V.: Informatics and Computer Technologies [in Ukrainian]. Naukova Dumka, Kyiv (2004)
164. Sergienko, I.V.: A method to solve problems of searching for extremum values. Avtomatika **5**, 15–21 (1964)
165. Mikhalevich, V.S., Kapitonova, Y.V., Letichevskii, A.A., et al.: System Software for the ES Multiprocessing Computer System [in Russian]. VVIA im. N. E. Zhukovskogo, Moscow (1986)
166. Statistical Yearbook of Ukraine for 2002 [in Ukrainian], Derzhkomstat, Kyiv (2003)
167. Stetsyuk, P.I.: r-algorithms and ellipsoids. Cybern. Syst. Anal. **32**(1), 93–110 (1996)
168. Tanaev, V.S., Shkurba, V.V.: An Introduction to Scheduling Theory [in Russian]. Nauka, Moscow (1975)
169. Trubin, V.A.: Network homogeneity deficiency and its reduction. Dokl. AN SSSR **318**(1), 43–47 (1991)

170. Trubin, V.A.: Dynamic decomposition method for linear programming problems with generalized upper bounds. Cybernetics **22**(4), 468–473 (1986)
171. Trubin, V.A.: A method to solve special ILP problems. Dokl. AN SSSR **189**(5), 552–554 (1969)
172. Trubin, V.A.: Integral packing of trees and branchings. Cybern. Syst. Anal. **31**(1), 21–24 (1995)
173. Trubin, V.A.: An efficient algorithm to solve arrangement problems on tree networks. Dokl. AN SSSR **231**(3), 547–550 (1976)
174. Uryas'ev, S.P.: Adaptive Algorithms of Stochastic Optimization and Game Theory [in Russian]. Nauka, Moscow (1990)
175. Khachiyan, L.G.: Polynomial algorithms in linear programming. Dokl. AN SSSR **244**(5), 1093–1096 (1979)
176. Khimich, A.N., Molchanov, I.N., Popov, A.V., Chistyakova, T.V., Yakovlev, M.F.: Parallel Algorithms to Solve Problems in Computational Mathematics [in Russian]. Naukova Dumka, Kyiv (2008)
177. Chikrii, A.A.: Conflict Controlled Processes [in Russian]. Naukova Dumka, Kyiv (1992)
178. Chikrii, A.A.: Multivalued mappings and their selectors in game control problems. Probl Upravl Informatiki **1**(2), 47–59 (1994)
179. Khimich, A.N., Molchanov, I.N., Mova, V.I., Perevozchikova, O.L., et al.: Numerical Software for the INPARKOM Intelligent MIMD-Computer [in Russian]. Naukova Dumka, Kyiv (2007)
180. Mikhalevich, V.S., Molchanov, I.N., Sergienko, I.V., et al.: Numerical Methods for the ES Multiprocessor Computer System [in Russian]. VVIA im. N. E. Zhukovskogo, Moscow (1986)
181. Shilo, V.P.: The method of global equilibrium search. Cybern. Syst. Anal. **35**(1), 68–74 (1999)
182. Shiryaev, A.N.: Sequential analysis and controlled random processes (discrete time). Kibernetika **3**, 1–24 (1965)
183. Shkurba, V.V., Podchasova, E.P., Pshichuk, F.N., Tur, L.P.: Scheduling Problems and Methods of their Solution [in Russian]. Naukova Dumka, Kyiv (1966)
184. Shor, N.Z.: Using space dilation operations in minimization problems for convex functions. Kibernetika **1**, 6–12 (1970)
185. Shor, N.Z.: Cut-off method with space dilation to solve convex programming problems. Kibernetika **1**, 94–95 (1977)
186. Shor, N.Z.: Methods for the Minimization of Nondifferentiable Functions and Their Applications [in Russian]. Naukova Dumka, Kyiv (1979)
187. Shor, N.Z.: Class of global minimum bounds of polynomial functions. Cybernetics **23**(6), 731–733 (1987)
188. Shor, N.Z.: An approach to obtaining global extremums in polynomial mathematical programming problems. Cybernetics **23**(5), 695–700 (1987)
189. Shor, N.Z.: Application of the gradient-descent method to solve the transportation network problem. In: Proceedingd of the Sci. Seminar on Theor. and Appl. Cybernetics and Operations Research, Issue 1, pp. 9–17. V. M. Glushkov Inst. of Cybernetics, AS USSR, Kyiv (1962)
190. Shor, N.Z., Zhurbenko, N.G.: Minimization method with space dilation toward the difference of two successive gradients. Kibernetika **3**, 51–59 (1971)
191. Shor, N.Z., Solomon, D.I.: Decomposition Methods in Linear Fractional Programming [in Russian]. Shtiintsa, Kishinev (1989)
192. Shor, N.Z., Sergienko, I.V., Shilo, V.P., et al.: Problems of Optimal Design of Reliable Networks [in Ukrainian]. Naukova Dumka, Kyiv (2005)
193. Shor, N.Z., Stetsenko, S.I.: Quadratic Extremum Problems and Nondifferentiable Optimization [in Russian]. Naukova Dumka, Kyiv (1989)

194. Podchasova, T.P., Portugal, V.M., Tatarov, V.A., Shkurba, V.V.: Heuristic Scheduling Methods [in Russian]. Tekhnika, Kyiv (1980)
195. Bignebat, C.: Labour market concentration and migration patterns in Russia. Marches Organization Institutions et Strategies d'Acteurs, Working Paper, No. 4 (2006)
196. Chikrii, A.A.: Game dynamic problems for systems with fractional derivatives. In: Chinchulun, A., Pardalos, P.M., Migdalas, A., Pitsoulis, L. (eds.) Pareto Optimality, Game Theory and Equilibria, vol. 17. Springer, New York (2008)
197. Chikrii, A.A.: Conflict Controlled Processes. Kluwer, Boston/London/Dordrecht (1997)
198. Ermoliev, Y., Wets, R.J.B. (eds.): Numerical techniques for stochastic optimization. Computational Mathematics, vol. 10. Springer, Berlin (1988)
199. Feo, T.A., Resende, M.G.C.: Greedy randomized adaptive search procedures. Glob. Optim. **6**, 109–133 (1989)
200. Gandolfo, G.: Economic Dynamic, Methods and Models. North-Holland Publ. Co., Amsterdam/New York/Oxford (1980)
201. Golodnikov, A.N., Knopov, P.S., Pardalos, P., Uryasev, S.P.: Optimization in the space of distribution functions and applications in the Bayes analysis. In: Uryasev, S. (ed.) Probabilistic Constrained Optimization: Methodology and Applications, pp. 102–131. Kluwer, Boston (2000)
202. Golodnikov, A.N., Knopov, P.S., Pepelyaev, V.A.: Estimation of reliability parameters under incomplete primary information. Theory. Decis. **57**, 331–344 (2005)
203. Ivanov, A.V.: Asymptotic Theory of Nonlinear Regression. Kluwer, Dordrecht (1997)
204. Kaniovski, Yu.M., King, A., Wets, R.J.-B.: Probabilistic bounds (via large deviations for the solutions of stochastic programming problems. Ann. Oper. Res. **56**, 189–208 (1995)
205. Kasyanov, P.O., Melnik, V.S., Yasinsky, V.V.: Evolution Inclusions and Enequalities in Banach Spaces with wλ-Pseudomonotone Maps. Naukova Dumka, Kyiv (2007)
206. Knopov, P.S., Kasitskaya, E.J.: Empirical Estimates in Stochastic Optimization and Identification. Kluwer, Dordrecht (2002)
207. Knopov, P.S., Pardalos, P.: Simulations and Optimization Methods in Risk and Reliability Theory. Nova Science Publishers, New York (2009)
208. Kuntsevich, V.M., Lychak, M.M.: Guaranteed Estimates, Adaptation and Robustness in Control Systems. Springer, Berlin (1992)
209. Lee, D., McDaniel, S.T.: Ocean acoustic propagation by finite difference method. Comput. Math. Appl. **14**, 305–423 (1987)
210. Lee, D., Pierse, A.D., Shang, E.C.: Parabolic equation development in the twentieth century. J. Comput. Acoust. **1**(4), 527–637 (2000)
211. Marti, K., Ermoliev, Y., Makowski, M., Pflug, G. (eds.): Coping with Uncertainty, Modeling and Policy Issues. Springer, Berlin (2006)
212. Marti, K., Ermoliev, Y., Pflug, G. (eds.): Dynamic Stochastic Optimization. Springer, Berlin/Heidelberg (2004)
213. Balinski, M.L., Wolfe, P. (eds.): Mathematical Programming, Study 3, Nondifferentiable Optimization. North-Holland Publ. Co, Amsterdam (1975)
214. Melnik, V.S., Zgurovsky, M.Z.: Nonlinear Analysis and Control of Physical Processes and Fields. Springer, Berlin/Heidelberg (2004)
215. Mikhalevich, M., Koshlai, L.: Modeling of multibranch competition in the labour market for countries in transition. In: Owsinski, J.W. (ed.) MODEST 2002: Transition and Transformation: Problems and Models, pp. 49–59. The Interfaces Institute, Warsaw (2002)
216. Mladenovic, N., Hansen, P.: Variable neighborhood search. Comput. Oper. Res. **24**(11), 1097–1100 (1997)
217. Chornei, R.K., Daduna, H., Knopov, P.S.: Control of Spatially Structured Random Processes and Random Fields with Applications. Springer, New York (2006)
218. Sergienko, I.V., Deineka, V.S.: Optimal Control of Distributed Systems with Conjugation Conditions. Kluwer, New York (2005)
219. Shor, N.Z.: Dual estimates in multiextremal problems. J. Glob. Optim. **2**, 411–418 (1992)

220. Shor, N.Z.: Nondifferentiable Optimization and Polynomial Problems. Kluwer, Boston/ Dordrecht/London (1998)
221. Shor, N.Z., Berezovski, O.A.: New algorithms for constructing optimal circumscribed and inscribed ellipsoids. Optim. Methods. Softw. **1**, 283–299 (1992)
222. Tappert, F.D., Lee, D.: A range refraction parabolic equation. J. Acoust. Soc. Amer. **76**, 1797–1803 (1984)
223. Glushkov, V.M., Ignatiev, M.V., Myasnikov, V.A., Torgashev, V.A.: Recursive machines and computing technologies. In: Proceedings of the IFIP Congress-74 (Stockholm), Amsterdam (1974)
224. Zhu, J., Lu, Y.Y.: Validity of one-way models in the weak range dependence limit. J. Comput. Acoust. **12**(1), 55–66 (2004)
225. Zgurovsky, M.Z., Pankratova, N.D.: System Analysis: Theory and Applications. Springer, Berlin (2007)

Index

I.V. Sergienko, *Methods of Optimization and Systems Analysis for Problems
of Transcomputational Complexity*, Springer Optimization and Its Applications 72,
DOI 10.1007/978-1-4614-4211-0, © Springer Science+Business Media New York 2012